Introduction to physical mathematics

Introduction to

physical mathematics

P.G. HARPER

Professor of Theoretical Physics, Heriot-Watt University, Edinburgh

D.L. WEAIRE

*Erasmus Smith's Professor, Department of Pure and Applied Physics
Trinity College, Dublin*

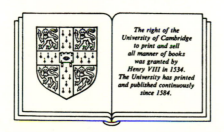

The right of the
University of Cambridge
to print and sell
all manner of books
was granted by
Henry VIII in 1534.
The University has printed
and published continuously
since 1584.

CAMBRIDGE UNIVERSITY PRESS

Cambridge

London New York New Rochelle

Melbourne Sydney

CAMBRIDGE UNIVERSITY PRESS
Cambridge, New York, Melbourne, Madrid, Cape Town, Singapore, São Paulo, Delhi

Cambridge University Press
The Edinburgh Building, Cambridge CB2 8RU, UK

Published in the United States of America by Cambridge University Press, New York

www.cambridge.org
Information on this title: www.cambridge.org/9780521262781

First published 1985
Re-issued in this digitally printed version 2008

A catalogue record for this publication is available from the British Library

Library of Congress Catalogue Card Number: 84–11411

ISBN 978-0-521-26278-1 hardback
ISBN 978-0-521-26908-7 paperback

Contents

Some elementary computer programs, together with answers to the exercises, are contained in a supplementary booklet. These may be obtained from the Department of Theoretical Physics, Heriot-Watt University, Riccarton, Edinburgh EH14 4AS.

Preface

The theoretical side of physical science holds up a mathematical mirror to nature. It seeks to find in the infinite variety of physical phenomena the few basic laws and relationships which underlie them. A secondary goal is the expression of these relations in efficient and transparent language.

After Newton had shown the power of this method, the eighteenth and nineteenth centuries saw its steady advance, hand-in-hand with experiment. At the end of the nineteenth century there was a crisis in physics – a widening gulf between theory and experiment – but, when Einstein emerged to resolve it, the new physics was still based on the old mathematics. It was simply used in surprising new ways. So it remains today, to a large extent, whatever educational theorists may tell us. Newton would not be greatly puzzled by the mathematics of Schrödinger's Equation.

On the other hand, the rapid development of computers is certainly changing our attitude to mathematics. This is obvious in the case of straightforward numerical calculations, but it extends also to the simulation of complex systems, the manipulation of algebra and even the proving of theorems. Applied mathematics is the art of the possible, and computers have widened its scope enormously. They are not just 'number-crunchers'. Nor are they available only to specialists. Most students today enjoy access to a powerful computer system, and many are skilled programmers at an early age.

Today's physical scientist needs both a feeling for the power of traditional analytic methods in relation to the physical world and an appreciation of modern computational methods. University curricula which rigidly separate mathematics, physics and computer science do not serve him well. At some stage, preferably at the beginning, these subjects should be brought together. This is what we have tried to do.

Our subject matter divides naturally into three parts. Elementary aspects of vectors, matrices and functions are introduced in chapters 1–11. The use of calculus and various approximate methods in solving physical problems, particularly those which involve differential equations, is covered in chapters 12–24. Chapters 25–34 introduce physical fields and the associated partial differential equations. Our objective throughout is the development of qualitative understanding and practical know-how, rather than rigour and completeness.

At various points in the text we have gone right back to basics, to explain things which university and college students have surely met before, such as differentiation. Most lecturers find that they must do this in an introductory course, whatever the curriculum may say, in order to bring all of the class up to a similar level of preparedness. Moreover, even the simplest operations raise many questions when we consider the pitfalls of their practical applications.

At Heriot-Watt, lectures based on this book run in parallel to a conventional course in pure mathematics (mostly calculus) in the first year, forming a bridge between the mathematics and physics courses. The class is given plenty of time to explore the exercises and encouraged to do so critically and creatively.

These exercises are indeed an integral part of the course and should be studied regularly. Some of the numerical ones are quite open-ended, since already students are using a great variety of computing hardware and it is hard to say what they may be using within a few years. Ideally, they should get some practice in the use of back-of-envelope arithmetic, hand calculators, and computers, including perhaps some library subroutines for numerical analysis. Some of the exercises make good material for a classroom discussion, reinforcing the message that applied mathematics is not a collection of cut-and-dried procedures but a flexible framework within which physical systems can be described in a variety of ways. This is not just an introduction – it is an *invitation*.

P.G. Harper
D.L. Weaire

Some notes on notation

The mathematical notation that we use is quite traditional. The following notes may be helpful in resolving some ambiguities.

Symbol   Meaning and Notes

→   *tends to*
The operation of taking a limit is the foundation of the calculus. The limit symbol should not be confused with $\sim$ (see below). Occasionally we also use an arrow to indicate replacement (so that $a \to b$ means 'a is replaced by b'), or displacement from one point to another.

$\sim$ *behaves as*
This always applies in some limit (sometimes implied rather than stated). For example, '$f(x) \sim g(x)$ as $x \to \infty$' means

$$\frac{f(x)}{g(x)} \to 1 \text{ as } x \to \infty.$$

$\approx$ *approximately equals*
Strictly speaking, this is rather meaningless unless there is some statement of the magnitude of the departure from equality (the error). Nevertheless it is widely used more casually to indicate the replacement of an exact value or formula by one which is not exact but of acceptable accuracy for the purpose at hand.

O ( )   *of order...*
Again a limit is involved. For example, $f(x) = O(x^3)$ as $x \to 0$ means $x^{-3} f(x) \to$ *a finite limit*, as $x \to 0$. It thus has a similar meaning to that of $\sim$, but less strict since the finite limit does not have to be unity. In particular, it may be zero.

Δ   *(Greek capital delta) increment of*
This is used here and in many introductory physics texts for an increment of a variable. Usually a *small* increment is implied, and often the limit is eventually taken in which all increments go to zero. This leads in some cases to differential relations, via

$$\frac{dy}{dx} = \lim_{\Delta x \to 0} \left(\frac{\Delta y}{\Delta x} \right)$$

and in others to integral relations.

v   *(bold typeface) a vector quantity*
A physical vector, or a column vector in matrix theory. In the latter, it is usual to reserve lower case letters for vectors, and capitals for matrices, except where tradition dictates otherwise!

$\overrightarrow{AB}$ *vector which represents the line AB*
In the use of vectors in association with geometrical constructions, this alternative notation denotes the vector of magnitude equal to the length of AB, in the direction indicated.

Σ   *(Greek capital sigma) 'sum'*
This has the usual meaning of a summation, but physical scientists often use it without the indication of the summation labels and range which are strictly required. Thus $\sum\limits_{i=1}^{N} m_i x_i$ might be written $\sum mx$ whenever its meaning is obvious from the context.

The Greek alphabet

In the interests of clarity and to avoid ambiguity, Greek symbols are often invoked in physical mathematics, to supplement the Roman alphabet. Many have traditional connotations – λ for wavelength, ν for frequency, ρ for density .... Mercifully, the use of other alphabets or typefaces, such as gothic, seems to be dying out.

	lower case	*capital*	
alpha	α	A	
beta	β	B	
gamma	γ	Γ	
delta	δ	Δ	(see notes above)
epsilon	ε	E	
zeta	ζ	Z	
eta	η	H	
theta	θ	Θ	
iota	ι	I	
kappa	κ	K	
lambda	λ	Λ	
mu	μ	M	
nu	ν	N	
xi	ξ	Ξ	
omicron	o	O	
pi	π	Π	
rho	ρ	P	
sigma	σ	Σ	(see notes above)
tau	τ	T	
upsilon	υ	Y	
phi	ϕ	Φ	
chi	χ	X	
psi	ψ	Ψ	
omega	ω	Ω	

The symbol ∇ is usually called 'grad' but, somewhat confusingly, ∇^2 is called 'del-squared'! These stand for operators (chapters 28, *et seq.*).

1
Introduction

Most physical laws express numerical relations between quantities which can be independently measured, such as the mass of a body, its acceleration and the force which is applied to it. Ultimately, they are established or refuted by experiment. The range of their validity is determined by the range of practicable experiments. Their generality is always in question, and physicists continually seek new insights in the breakdown of old theories.

It is important therefore to distinguish between physical *laws*, which are provisional and approximate (since, in principle, we expect to find circumstances in which they do not apply), and other mathematical relationships which are merely conventional *definitions*, such as 'momentum equals mass times velocity'. These cannot be overturned, although there may be a time or a place in which they are not useful.

Given a problem to solve, we make the transition to mathematics by choosing appropriate physical laws and definitions. For the purposes of mathematical manipulation we may provisionally regard this formulation as exact, but in practice we will soon encounter uncertainties of two kinds.

First, if we wish to use experimentally measured quantities as numerical input to our calculations, as must ultimately be the case, we should recognise that every measurement involves some degree of uncertainty. The word 'error' is commonly used for this, which is unfortunate, because it need not be the result of any mistake or misjudgement, but may simply follow from the limited accuracy of the available measuring apparatus. We should be aware of the magnitude of this error and try to trace its effects through our calculation, to see what bearing it has on the final result or output. This can then be assigned some estimate of uncertainty or 'error bar'.

Secondly, any numerical calculation that we perform entails further errors. These may arise from the round-off error of the computer or calculator (since only a finite number of digits can be retained), or the approximations of the numerical methods which are used.

In chapter 2 we shall look at errors in more detail. In the rest of the chapters we shall talk about them only when some interesting point arises, e.g. when there is a possibility of large errors in a particular numerical method. However, in every real application of mathematics this aspect must not be overlooked.

It must also be borne in mind that physical quantities are expressed in terms of units, and only rarely do we meet dimensionless quantities which do not require them. Even in a mathematically oriented course units must be respected.

On the other hand, whenever one is concentrating on the mathematical aspects of a theory the units may often be disregarded, as we do from time to time. The closer one gets to real applications the more important it is to remember that all input values and all final results should have clearly stated units. If one works within a consistent scheme of units such as the SI system one can often ignore units at intermediate stages in a calculation, and assign the obvious units to the final result.

So the full result of a measurement or calculation should look something like the following (for an acceleration):

$$a = 6.32 \pm 0.02 \, \text{ms}^{-2}. \tag{1.1}$$

Without some reference to units, this would be meaningless. Without the error estimate ± 0.02, it would still be acceptable, but would carry the implication that the result 6.32 was expected to be correct to within 0.005, i.e. that errors are unlikely to affect the last digit which is retained. In either case, some thought must be given to the number of significant figures which are retained. The above might have been the result of a computer calculation which printed out 6.322 371. The error estimate causes us to round off this number by the elimination of the last four digits, because they are insignificant. It would have been pointless to include more digits, in view of the error associated with the result. Indeed if the error had not been stated, extra digits would have been quite misleading. People often forget to 'prune' their results in this way. An amusing example is provided by the Laws of Rugby Football in which, for the metrically minded referee, it is recommended that the ball should have 'a pressure equivalent to 0.6697–.7031 kilograms per square centimetre at sea level'. Some worthy official worked this out by converting the old fashioned '9–10 pounds per square inch', but would have done well to drop at least two figures from his results.

What about the actual mathematical methods – how do they proceed? You should try to develop a flexible, balanced attitude to this, and remember that there are at least two goals – quantitative description and qualitative understanding. Prescribed questions for homework and examinations may call for some obvious technique which you have just learned. But in reality, any given problem in physical mathematics can be attacked from many directions. Are there powerful theorems which will transform or reduce it to something simpler? Is there an 'exact' analytical solution? (We shall discuss the meaning of this in chapter 10.) Is there an exact solution to a related problem, which can be used as a starting approximation? Having exhausted this line of enquiry, the applied mathematician nowadays takes his problem to the computer, in the form most amenable to calculation. Again, there may be a choice of methods. Increasingly, this is done at quite an early stage because today's machines can do wonders with even clumsy and elementary numerical methods. Often, the results of this 'brute force' approach will inspire analytical methods or approximations that could not be formulated in the first place. Of course, such a comprehensive view cannot be acquired all at once, but at every stage you should try to 'look over your shoulder' at methods you have already studied which relate to the subject at hand.

While exceptional individuals are completely comfortable with symbolic mathematics, most of us make visual images of even the most abstract relationships. Sketches are always helpful in this respect. We have included many of these, and the student is encouraged to do likewise. Computer graphics have begun to be very widely used as an aid to evaluating numerical output. Increasingly, moving images and stereographic projections are used. We have taken a very modest step in this exciting direction by including computer-generated graphical output wherever appropriate. Much of this was produced by the same few simple commands to a software package associated with an HP7470A Plotter, so similar plots can be easily generated and extended by students or lecturers. In the absence of a plotter, a simple 'line-printer plot' will often suffice.

Summary

1.1. Flow chart for practical mathematicians.

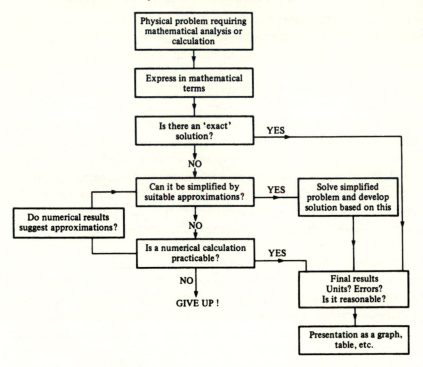

EXERCISES

1. Explain in about a paragraph the meaning of each of the following terms used in the text: *round-off, analytical, stereographic, software, qualitative, line-printer plot.*

2. Discuss the following assertion, taken from the preface: 'applied mathematics is the art of the possible'.

3. The population of a country is an integer with discontinuous changes, yet it is often described as if it were a continuous function of time. Discuss this, contrasting it with the case of the size of a family.

2
Errors

Physical scientists use a lot of common sense and only a little statistics in dealing with errors. Only in special areas (such as the work of Standards Laboratories) are errors relentlessly pursued with full rigour. Usually, it is enough to be satisfied that they are negligible or insignificant for the purposes at hand. As with safety regulations, one tries to ensure a large 'margin of error' wherever possible. This contrasts with the state of affairs in biology and in social sciences. These deal in quantities, such as the frequency of the human heartbeat, so widely variable that errors must be treated with caution and proper statistical methodology. But even common sense needs a little mathematical background, such as is given here.

First, we may distinguish between *random* and *systematic* errors. A systematic error is one which is consistent from one measurement to the next. It might arise from inaccurate adjustment of instruments, a faulty calibration, the ineptitude of the scientist himself, or simply from a failure to recognise some influence upon the data which was not the object of the experiment. We can try to identify such errors and either eliminate them or add corrections to allow for them, whereupon they no longer concern us. It is not usual to include systematic errors in an error estimate since, if we can identify them, they can be removed! Hence we shall concentrate on the second type of error, which is random, i.e. the error associated with one measurement has nothing to do with the next or subsequent measurements. This kind of uncertainty, which is due to some fluctuating influence upon the measurement, is recognised if we repeat one measurement several times and obtain slightly different results. Unlike the systematic error, random error should not affect the *average* of a sufficiently large number of measurements. This is the key to its reduction. How then are we to do this and how are we to express the final result?

To make this concrete, consider the measurement of the diameter of a standard pencil taken from stock. A micrometer can apparently measure this to within a few per cent of a millimetre. But in applying the gauge to the soft wood, we may compress it slightly. To the extent that we do this consistently, the error is systematic and could be identified by suitable experiments and so allowed for. There would, however, remain a variable part of the error, due to a variable force being applied to tighten the

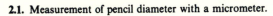

2.1. Measurement of pencil diameter with a micrometer.

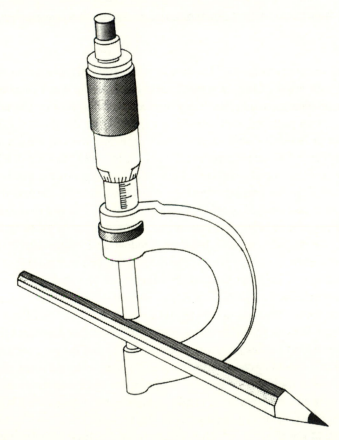

micrometer. This might be assumed to be random. If this uncertainty is greater than that associated with the instrument itself and its scale, it must take over as the main source of the error associated with the measurement.

The pencil width itself will still be variable. This could be explored simply by repeating the measurement on different parts of a single pencil, or on different pencils. In fundamental physics, life is simpler—it seems that every two electrons, for instance, are exactly the same, as far as we can tell. The only uncertainty in our knowledge of the mass or charge of an electron is that derived from the process of measurement. However, with pencils as with much else, we encounter the kind of intrinsic variability which becomes so serious in biological science. It will make our arguments more concrete if we concentrate upon this variability, supposing it to be much greater than the uncertainty of measurement.

Our procedure will then be to measure the diameters of a large number of individual pencils, and try to extract from these a standard value and uncertainty limits, rather as we did in eq. (1.1). What do we mean by a standard value? It is natural to take the average or mean value from our measurements as an estimate of the mean value $\langle x \rangle$ which would be obtained in the limit of an infinite number of such measurements. We can get an immediate feeling for the random deviations from the average value by plotting the distribution of measured diameters, as in fig. 2.2. Ideally, this is a

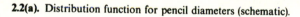

2.2(a). Distribution function for pencil diameters (schematic).

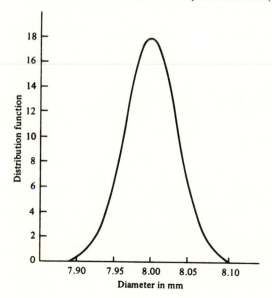

continuous function (fig. 2.2(*a*)) whose definition is

fraction of pencils with diameters
between x and $x + \Delta x$
$$= f(x)\Delta x \tag{2.1}$$

in the limit of small Δx. With a finite number of measurements, we may choose to represent them in a histogram (fig. 2.2(*b*)), in order to identify the shape of the function f. Each part of the histogram gives the fraction of pencils in our sample which lie in a chosen range. With only a finite sample, we have to choose wide enough ranges in our histogram, for each to contain a reasonably large number of pencils.

The function f may be described as giving the relative *frequency* with which randomly chosen pencils will be found to have diameters close to a given value. The word *probability* may also be used. This has wider and more controversial meanings outside physical science – what is the probability that Fool's Gold will win the 3.30 at Newmarket? As John Locke said in 1690: 'probability is either Matter of Fact, or Speculation'. Happily, the wider meanings of probability need not concern us here.

Such a distribution function need not have any particular shape. In some cases (although hardly with our pencils) it stretches all the way between plus and minus infinity. Even if this is not the case, an error estimate does not usually bracket the entire distribution. Rather than trying to say that x

2.2(b). Histogram representation of the distribution function, prepared from sample measurements.

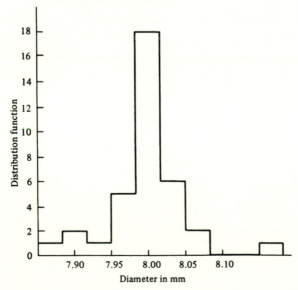

must lie between x_1 and x_2, we usually say or imply that this is very probable.

Often the distribution function $f(x)$ has a smooth, symmetric, bell-like shape. There are good reasons for this, which suggest the Gaussian function (chapters 10, 25 and exercise 3) as the appropriate function in many cases.

Let us return to our original goal – to establish a standard diameter, which we take to be the average value, for pencils of a given type. We now see that if we were to take just one measurement as an indication of this, the uncertainty would be dictated by the width (in some sense) of the peak of the curve in fig. 2.2(a), but usually we would not know this in advance. If we measured a lot of individual pencils, we could construct Fig. 2.2(a), and estimate the uncertainty from this, as well as taking the average. But clearly the uncertainty of our calculated average is no longer that of an individual measurement – it must decrease continually as we take more and more pencils into our sample.

There is a simple rule for this, as follows:

uncertainty of average of x, estimated from N measurements

$$= \frac{\text{uncertainty for one measurement}}{N^{\frac{1}{2}}}. \qquad (2.2)$$

Thus a sample of 100 pencils is ten times better than one pencil, as far as the uncertainty associated with an estimate of the average diameter is concerned.

Let us try to understand how this comes about and, in due course, justify it more precisely. It is clear at the outset that it has something to do with the cancellation of the deviations from $\langle x \rangle$ associated with individual pencils, which can be positive or negative. To simplify analysis, let us suppose that the diameter x deviates randomly from its average value $\langle x \rangle$ by precisely $\pm \varepsilon$, with equal probabilities, i.e. diameters of $x + \varepsilon$ are just as common as diameters of $x - \varepsilon$,

$$x = \langle x \rangle \pm \varepsilon. \qquad (2.3)$$

What we want to do is to estimate the uncertainty associated with an estimate of $\langle x \rangle$ from ten pencils. This might be made conveniently by laying them side-by-side and measuring their combined width X (see exercise 5).

The combined width X of ten pencils placed side-by-side could be as big as $10\langle x \rangle + 10\varepsilon$ (all positive deviations), or as little as $10\langle x \rangle - 10\varepsilon$ (all negative deviations), but is more likely to take some intermediate value (both positive and negative deviations). By measuring a succession of batches of ten, all possible values of X would eventually be met. Any particular value X will occur with a frequency proportional to its weight, defined to be the number of equivalent ways in which it could be realised, since each of these is equally probable. The maximum width, $X = 10\langle x \rangle + 10\varepsilon$, can be formed in only one way (all deviations positive, i.e. the arrangement $+ + + + + + + + + +$). The next lowest value, $X = 10\langle x \rangle + 8\varepsilon$ could be realised by the arrangements $- + + + + + + + + +, + - + + + + + + + +$, etc., and therefore has the weight 10. The value $X = 10\langle x \rangle + 6\varepsilon$ (8 positive, 2 negative deviations) could be made up in 45 ways. Each of the ten arrangements like $+ + - + + + + + + +$ could be converted to one like $+ + - + + + + - + +$ by reversing just one sign, to give nine possibilities. The resulting arrangements, however, duplicate each possibility since reversing number 3 first, and number 8 second gives the same as reversing number 8 first and number 3 second. Thus the weight is not 9×10 but $9 \times 10/2 = 45$.

Proceeding in this way to 3, 4 and 5 reversals, we may tabulate our results as in table 2.1. See also fig. 2.3. All of these results are expressible in terms of factorials. The weight for deviation $\pm n\varepsilon$ is $10![n!(10 - n)!]^{-1}$ and students familiar with combinatorials may be able to jump to this result. (See exercise 2 for the use of combinatorials.)

The total number of possible arrangements (sum of weights) is 1024. This is just 2^{10}, because there are two possibilities for each of ten pencils. On dividing each weight by 1024 the resulting fraction is the probability of the corresponding deviation. Clearly the sum of the probabilities is one. The most likely total deviation is zero, with a probability of $252/1024 = 0.246$.

With the aid of this table of probabilities, the earlier questions can now be answered. Multiply each value of X by its probability, i.e. $X = 10\langle x \rangle + 10\varepsilon$ by 0.001, $X = 10\langle x \rangle + 8\varepsilon$ by 0.01 etc. and add. Because of cancellation, $(10\varepsilon$ with $- 10\varepsilon$ etc.) and because the sum of the probabilities is 1.0, the mean

Table 2.1   *The addition of ten numbers randomly chosen as $\langle x \rangle - \varepsilon$ or $\langle x \rangle + \varepsilon$ gives the result X with probabilities for the various possible outcomes as shown.*

$X - 10\langle x \rangle$ (deviation from mean)	$\pm 10\varepsilon$	$\pm 8\varepsilon$	$\pm 6\varepsilon$	$\pm 4\varepsilon$	$\pm 2\varepsilon$	$+ 0$
Weight	1	10	45	120	210	252
Probability = weight/1024	0.001	0.010	0.044	0.117	0.205	0.246

value of X is given by

$$\langle X \rangle = 10\langle x \rangle \tag{2.4}$$

as expected. This simply says that we could get the same final average diameter in the limit of a large sample if we grouped our pencils in tens, as if we measured them individually.

However, what we are really looking for is the uncertainty associated with the average taken from only ten pencils. To estimate the spread of values about the mean, statisticians use the mean square deviation, or its square root (rms). By squaring the deviations, we ensure that cancellation cannot occur. Thus multiplying each value of $(X - 10\langle x \rangle)^2$, namely, $(10\varepsilon)^2$, $(-10\varepsilon)^2$, $(8\varepsilon)^2$ etc., by the appropriate probability (taken from the table) and adding, the sum becomes

$$2(1 \times 10^2\varepsilon^2 + 10 \times 8^2\varepsilon^2 + \cdots)/1024.$$

Using the symbol ΔX to denote the deviation of X from $10\langle x \rangle$, the relation is written as a mean value,

$$\langle (\Delta X)^2 \rangle = 10\varepsilon^2.$$

Its square root is the rms deviation of x, but what we really want is the rms deviation of $X/10$, which is $10^{-\frac{1}{2}}\varepsilon$. Note that, for our simple distribution, ε is the rms deviation for a single measurement. So we have seen that in this

2.3. Plot of data from Table 2.1, giving the probability of deviations from the mean.

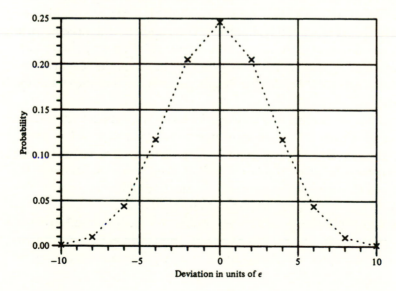

specific case the uncertainty (in the sense of the rms deviation) has been reduced in accordance with eq. (2.2).

This important formula can be derived more generally as follows, if one understands how to manipulate probabilities. Suppose that for N measured values, the deviations are

$$\Delta x_1, \Delta x_2, \Delta x_3, \Delta x_4, \ldots, \Delta x_N.$$

Then the deviation of their sum X is given by the sum

$$(\Delta X)^2 = (\Delta x_1)^2 + (\Delta x_2)^2 + \cdots + (\Delta x_N)^2$$
$$+ 2\Delta x_1 \Delta x_2 + 2\Delta x_1 \Delta x_3 + \cdots . \tag{2.5}$$

The trick is to recognise what happens if we take an average of both sides. Each of the squares $(\Delta x_1)^2$ has the same average value, namely $\langle (\Delta x)^2 \rangle$, and there are N of them. But the products such as $\Delta x_1 \Delta x_2$ can be positive or negative, and in each case the average is zero. Thus

$$\langle (\Delta X)^2 \rangle = N \langle (\Delta x)^2 \rangle . \tag{2.6}$$

This formula agrees with the earlier expression, taking $N = 10$.

Remember that we must divide the sum of our measurements by N to obtain the average, so the root mean square deviation of this *decreases as* $N^{\frac{1}{2}}$ as N increases:

$$\langle (\Delta X/N)^2 \rangle^{\frac{1}{2}} = \langle (\Delta x)^2 \rangle^{\frac{1}{2}}/N^{\frac{1}{2}}. \tag{2.7}$$

The rms deviation is often used as the estimate of the uncertainty of the average, i.e. we calculate this quantity as well as the average of our set of measurements and use it as our estimate of the 'half-width' of a distribution $f(x)$. A further example of this important $N^{\frac{1}{2}}$ factor in operation is shown in fig. 2.4, which shows the variation of the estimate of the average taken from a sequence of random numbers between 0 and 1, as the length of the sequence is extended. For further details, see exercise 4 at the end of the chapter.

The above discussion was couched in terms appropriate to random errors arising from the variation of individual specimens, but it clearly applies just as well to the random errors of the measurement itself. If the two (or any other *independent* sources of random error) are to be combined the rule is: *add the mean square deviations* arising from different sources. This can be proved by methods similar to those used above.

The same rule can generally be applied to the round-off error in numerical calculations, which is rather like measurement error. In particular, note the danger of subtracting one number from another, if they are nearly equal. The absolute value of the uncertainty is increased by only $2^{\frac{1}{2}}$ but it is greatly increased relative to the difference of the numbers, and it is *relative* uncertainty that usually matters.

What happens if we multiply (or divide) numbers? The rule in this case is that the uncertainties can be added only after being expressed as *fractions* of the quantities involved. Essentially this follows from

$$\frac{(x + \Delta x)(y + \Delta y)}{xy} \approx 1 + \frac{\Delta x}{x} + \frac{\Delta y}{y}$$

and it is valid provided the uncertainties are small compared with the quantities themselves.

In practical numerical calculation it is usually difficult to trace the uncertainty through the many steps involved, and errors due to approximations (which may be systematic!) are also involved. Often, confidence in the final result is based not on any formal error analysis, but rather on simple tests of the effect of the approximations and round-off error in exploratory calculations. One can, for instance, repeat a computer calculation in 'double precision' to greatly reduce round-off error, and see what happens. Also, there may be available some established, accurate solution to a particular version of the problem which can be used for test purposes.

2.4. A random number generator was used to calculate the average of a sequence of N random numbers uniformly distributed between 0 and 1. As N is increased, this average approaches 0.5 as shown.

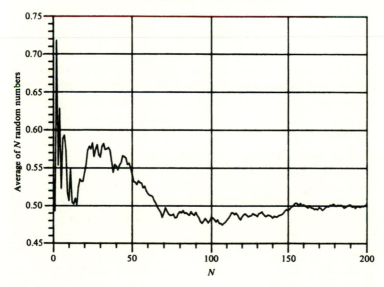

Summary

The uncertainty in an individual value due to random errors is conveniently characterised by the rms deviation from the average value $\langle x \rangle$. The accumulated uncertainty in the sum of N such values is expressed through the mean square deviation $\langle (\Delta X)^2 \rangle$ and given by

$$\langle (\Delta X)^2 \rangle = N \langle (\Delta x)^2 \rangle.$$

This means that the uncertainty of the estimate of the average of x from N measurements is $\pm \langle (\Delta x)^2 \rangle^{\frac{1}{2}} / N^{\frac{1}{2}}$.

Mean square deviations from independent sources are to be added. When quantities are multiplied or divided, *fractional* uncertainties are to be added.

EXERCISES

1. For the pencil-width problem discussed in the text, calculate the total probability that the deviation of the width of a batch of ten pencils from its mean value will be 4ε or more. (Use the table or graph provided.) If 200 batches are examined, how many would be expected to have such deviation?

2. If the letters of the word PARABOLA are re-arranged at random, what is the probability that the word BAR will somewhere occur?

3. The *normal* probability distribution function of x-values is

$$f(x) = \left(\frac{2}{\pi\sigma} \right)^{\frac{1}{2}} \exp \left(-\frac{x^2}{2\sigma^2} \right).$$

   When written in this way, it has the mean $\langle x \rangle = 0$, and σ^2 is the mean square deviation (see text). Such a function is called a Gaussian function. Calculate $f(x)$ for integer x-values in the interval $-10 \leqslant x \leqslant 10$, taking $\sigma^2 = 10$ and check that it is a good approximation to tabulated values in the text. This correspondence becomes exact in the limit of large 'batches' (in the sense used in the text). The Gaussian function emerges as the correct form whenever a large number of different sources of error are combined additively.

4. Most computers have a 'random number generator' which may be used to generate numbers with *uniform* probability between 0 and 1. Sketch the corresponding distribution function. Use such a generator (or a table of random numbers, to be found in any library), to repeat the calculation shown in fig. 2.4. Repeat the calculation for a different set of random numbers. Discuss the results. Give a *sketch* of the shape that you would expect the distribution function to have for the sum of two such numbers.

5. The measurement of the total width of N pencils could, of course, be made by laying them side-by-side and making just one measurement. This

method, applied to people's feet (laid end-to-end) was advocated by a German author in the sixteenth century, in trying to define the (average) 'foot' as a unit of measurement (fig. 2.5). Discuss the advantages and disadvantages of this procedure.

6. By how many significant figures would the accuracy of the calculation of
   (22/7 − 355/113)
   be reduced by round-off error, if it is calculated in the obvious straightforward way? How could the calculation be re-organised to avoid this?

7. Three different methods of representing a distribution function graphically have been used in this chapter, in figs. 2.2(*a*), 2.2(*b*) and 2.3. Discuss this.

8. A class of students assigned an experiment to measure the gravitational constant g using a pendulum provided the following experimental values (in ms^{-2}):
   9.84, 9.20, 9.87, 9.99, 10.07, 10.01, 9.78, 9.85, 9.67, 10.30, 9.79, 9.95, 9.98, 9.86, 9.87, 9.93, 10.94, 9.88, 9.83, 10.1, 10.38, 9.65, 9.39, 9.51, 9.99, 10.02, 9.87, 9.39, 9.87, 9.87, 10.20, 9.65, 10.12, 9.86, 9.99, 9.87.
   (*a*) Construct a histogram to show the distribution of values.
   (*b*) Find the mean value and rms deviation.
   (*c*) What error would you associate with this value of g, disregarding systematic error?

2.5. Early German definition of the British foot!

(*d*) Comment on the sources of random and systematic error in this experiment.

9. Estimate the probability that, in your class, two (or more) students have a birthday in common. Assume an equal probability of all dates (is this correct?) and a randomly chosen class. (*Hint*: the probability that a named student has a birthday on a given date is roughly 1/365, so that the probability for two named students to have specified birthdays is $(1/365)^2$. But if we don't name the students, and don't name the date, we increase the probability.)

10. A standard brick of length 8.5 in. has an intrinsic uncertainty in length of 0.1 in. What length of bricks laid end-to-end would be such that an estimate of the number of bricks, based on this value, would have an uncertainty of *one*?

3

Cartesian coordinates

The non-mathematical New Yorker has no difficulty in locating places – he uses Cartesian coordinates. Numbered streets run east–west and avenues run north–south. If he lives in a city apartment, his floor provides the third coordinate. For the physicist or chemist, atoms in molecules in space are locatable with a similar grid system. Three directions or axes marked out in blocks or length units can provide the unambiguous address of any chosen atom or event. Since the axes, unlike the New York roadways, are fictional, there is some freedom of choice, and this has mathematical consequences. Although any set of axes (mutually perpendicular) will do, if there are other physically significant directions already present, such as the direction of gravity, it generally pays to use them. The house numbers, so to speak, can be counted in either direction so that, having chosen 'zero', both positive and negative locations may appear. The use of such coordinates was first introduced by the seventeenth-century French philosopher and scientist René Descartes.

In one dimension (in which we never depart from a straight line), all positions may be measured as displacements (in a chosen direction) from a fixed point, the origin, and are generally denoted by x. So x is continuous and can be positive or negative. Two points x_1, x_2 have a separation $|x_2 - x_1|$ (always positive), and the displacement of x_2 from x_1 is $x_2 - x_1$ (may be positive or negative). Successive displacements $x_1 \rightarrow x_2$ $\rightarrow x_3$ etc. can be added (with either sign) to give a net displacement,

$(x_2 - x_1) + (x_3 - x_2) + (x_4 - x_3) = x_4 - x_1$. The order does not matter: $(x_2 - x_1) + (x_4 - x_3) + (x_3 - x_2)$ still equals $x_4 - x_1$.

An example from physical chemistry illustrates the use of such one-dimensional coordinates. The problem is to locate the centre of mass of the molecule iodine cyanide, ICN, whose three atoms lie in a straight line with the carbon atom lying between the iodine (I) and nitrogen (N) atoms. The I–C bond length is known to be 0.199 nm, while the C–N bond length is 0.116 nm. The coordinate origin may be taken anywhere, but choosing it at the C position, with an x-axis in the I–C direction, the atomic coordinates and approximate atomic weights are shown in table 3.1. The formula for the centre of mass (centroid) coordinate $\bar{x}$ is $M\bar{x} = m_1 x_1 + m_2 x_2 + m_3 x_3$ where M is the molecular mass and $M = m_1 + m_2 + m_3$. Using the table values, $\bar{x} = -0.155$ nm. It is worthwhile repeating the calculation taking the origin at the I atom and checking that the resulting x gives the same position relative to the molecule.

Few physical situations can be adequately described with only one dimension. At least a second (perpendicular) axis is needed, with an independent coordinate generally denoted by y. The origin for y-measurements is on the x-axis. Figure 3.1 makes it clear that a point P is reached by an x- and a y-displacement and that the order does not matter. The directions associated with x and y define a plane, usually

Table 3.1 *Atomic coordinates and masses for the molecule ICN*

Atom	Coordinates	Atomic weight
I	$x_1 = -0.199$ nm	$m_1 = 126.9$
C	$x_2 = 0$	$m_2 = 12.0$
N	$x_3 = 0.116$ nm	$m_3 = 14.0$

3.1. Coordinates of a point P in two dimensions.

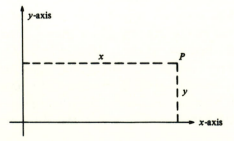

referred to as the (x, y) plane. Every point in this plane can be located by its coordinates, namely the pair (x, y). Using the theorem of Pythagoras, the separation of the two points (x_1, y_1) and (x_2, y_2) is $[(x_1 - x_2)^2 + (y_1 - y_2)^2]^{\frac{1}{2}}$ and this is always positive. The displacement of x_2, y_2 from x_1, y_1 is the pair $(x_2 - x_1, y_2 - y_1)$ and the components $x_2 - x_1$ and $y_2 - y_1$ may be of either sign. Figure 3.2 depicts some possible arrangements.

As an example, suppose a point is displaced from its original position $(6,1)$ by an unknown distance in the x-direction, and by -4 units in the y-direction. If it is known that the distance from the origin remains the same, what is the x-displacement? Draw the figure, fig. 3.3, and set it up as follows. Original point, P, coordinates $(6, 1)$; displaced point, Q, coordinates $(x_2, 1 + (-4))$, x_2 unknown. Then $OP = (1^2 + 6^2)^{\frac{1}{2}} = OQ = (x_2^2 + 3^2)^{\frac{1}{2}}$. This last relation provides an equation to find x_2, namely, squaring both sides, $37 = x_2^2 + 9$, whence it follows that $x_2 = 28^{\frac{1}{2}} = \pm 5.29$. The displacement is therefore $x_2 - 6 = -0.71$ or -11.29. Both values are acceptable, and it is worthwhile completing fig. 3.3 to show them.

3.2. Positive/negative Cartesian coordinates
$x_1 \langle 0, y_1 \rangle 0 \, ; \, x_2 \rangle 0, y_2 \langle 0 \, ; \, x_2 - x_1 \rangle 0, y_2 - y_1 \langle 0$

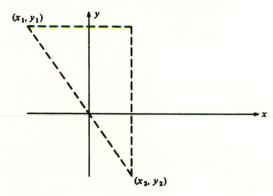

3.3. Graphical solution of the example in the text.

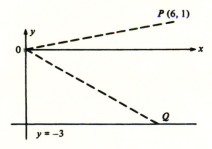

Two dimensions are sufficient to describe many physical situations. Often, however, a third dimension is needed, specified by a z-axis perpendicular to x- and y-axes, i.e. perpendicular to the (x, y) plane. The z-origin is located in this plane so that a point is completely specified by coordinates (x, y, z). This clearly suffices to describe the geometry of three-dimensional space, but abstract spaces may have more dimensions. For example, four-dimensional space is useful in relativity.

The separation of the two points (x_1, y_1, z_1), (x_2, y_2, z_2) in three-dimensional space is given by $[(x_2 - x_1)^2 + (y_2 - y_1)^2 + (z_2 - z_1)^2]^{\frac{1}{2}}$ and the displacement (or relative location) of (x_2, y_2, z_2) with respect to (x_1, y_1, z_1) is $(x_2 - x_1, y_2 - y_1, z_2 - z_1)$.

For example, three-dimensional coordinates may be used to describe the atomic positions of atoms in molecular structures. In the ammonia molecule NH_3, the three hydrogen atoms form an equilateral triangle, with the nitrogen equidistant from each corner, but not in the same plane. The

3.4. Atomic configuration of the ammonia molecule NH_3

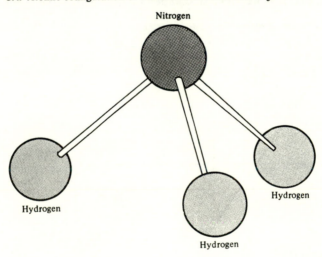

3.5. The hydrogen atoms of the NH_3 molecule are arranged in a plane as shown.

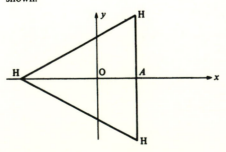

arrangement is shown in the perspective drawing, fig. 3.4.

What are the Cartesian coordinates of each of the four atoms? It all depends where the origin is placed and what directions are chosen for x, y, z. The plane containing the hydrogen atoms could serve as the (x, y) plane say, and then obviously a perpendicular axis, z, could go through the nitrogen atom. The simplest choice for the y-axis would be parallel to the line through two hydrogen atoms, as in fig. 3.5.

Suppose the H–H separation is l, and $OA = x$. The y-coordinates of two of the hydrogen atoms are simply $\pm \frac{1}{2}l$. The remaining hydrogen has a negative x-coordinate, given by $x - l = -l/3^{\frac{1}{2}}$. Of course the z-coordinates of all the hydrogens are zero. Finally the nitrogen atom is on the z-axis so that its (x, y) coordinates are zero. Placing the origin in the plane of the hydrogen atoms makes for simplicity, but in dealing with rotation a better choice would be the centre of mass of the molecule. This clearly lies on the same z-axis (through the N atom) but divides the apex height h in the ratio 3:14, taking the atomic weights of N and H as 14 and 1. Referred to this origin, the N coordinate becomes $(0, 0, \frac{3}{17}h)$.

Equipped with three-dimensional coordinates, the centroid formula given above can be extended by adding similar equations for $\bar{y}$ and $\bar{z}$. The equations are quite independent so that, for instance, re-arranging the masses within the (x, y) plane would alter $\bar{x}$ and $\bar{y}$ but would leave $\bar{z}$ unaltered.

If the individual atoms carry electric charges and the total is zero (which is quite often the case) we can also calculate three quantities which together constitute the electric *dipole moment* of the molecule. These are defined by $\mu_x = e_1 x_1 + e_1 x_1 + e_2 x_2 + \cdots$, with a term for each charge e_i, and two more equations for the y and z directions. So long as the molecule is electrically neutral, μ_x, μ_y and μ_z are independent of the choice of origin. Their significance will emerge in later chapters.

Summary

Location or position is generally specified by a set of coordinates x, y, z taken along three mutually perpendicular directions, and measured from the (y, z), (z, x) and (x, y) planes respectively.

Displacements $x_1 \to x_2$, $y_1 \to y_2$, $z_1 \to z_2$ can be taken in any order, and are additive.

Separation of, or distance between, points (x_1, y_1, z_1) and (x_2, y_2, z_2) is $[(x_1 - x_2)^2 + (y_1 - y_2)^2 + (z_1 - z_2)^2]^{\frac{1}{2}}$.

EXERCISES

1. A displacement $(0, 4, -1)$ is followed by a second displacement to give a net displacement $(6, 6, -8)$. Find the second displacement.

2. Successive displacements are as follows: $(0, 1)$, $(2, -3)$, $(-3, 4)$ and $(4, 1)$. Plot these on squared paper and measure the total displacement from the origin. Confirm by calculation.

3. A displacement whose length is 3.5 at an angle of $20°$ to the x-axis is followed by a displacement of length 2.4 at an angle of $-40°$ to the x-axis. Express these displacements in terms of x, y coordinates, and find graphically and by calculation the net displacement and separation from the origin.

4. Two displacements of unknown length, one at $30°$ to the x-axis, the other at $-60°$ to the axis, combine to give the net displacement $(3.16, 0.73)$. Find the unknown lengths.

5. Find the coordinate of the point which lies on the line joining x_1 to x_2, and divides their separation in the ratio $2:1$.

6. Find the coordinates of the centroid of three masses, 1 kg at $(1, 0)$, 1 kg at $(0, 1)$ and 4 kg at $(5, 5)$, and re-express these positions relative to the centroid as origin.

7. The atoms of the water molecule H_2O form a triangle with the bond angle H–O–H equal to $104.45°$. Taking the atomic weights of H and O as 1.0 and 16.0, and units such that the bond length is unity, find the coordinates of each atom referred to an origin at the centroid, and x-axis parallel to the H–H axis.

8. Two points A, B have coordinates $(1, -1, 4)$, $(2, 3, 1)$. A third point C has coordinates $(x, y, 0)$. Find x, y if A, B, C lie in a straight line.

9. When drilling bore-holes, the bit commonly deviates from the vertical by an angle θ in a vertical plane inclined at an angle ϕ to the north, say. Readings from a gyro-compass at successive drill 'strings' of length 10 m are as follows:
θ: $1°$ $3°$ $4°$ $10°$ $15°$ $15°$ $15°$
ϕ: $0°$ $10°$ $12°$ $20°$ $35°$ $68°$ $85°$.

Taking the z-direction as vertical and the y-direction as north, find the x, y coordinates of successive positions of the drill bit.

10. There is a further choice to be made in defining Cartesian axes, not mentioned in the text. Given x- and y-axes there are two possibilities for the z-axis. In practice, a 'right-handed' set is conventionally chosen and this will be defined in chapter 6. Can you think of any definition which would specify one of the two possibilities, without reference to some object such as a right hand, doorknob, screwdriver or whatever?

4

Vectors

The latest VTOL (vertical take-off and landing) aircraft are described in the press as having 'vectored thrust'. The jet can indeed be rotated relative to the aircraft so as to produce an orientated propelling force. In less newsworthy applications, physicists and engineers have been dealing with vectored thrust for a century or so, and in an immense variety of situations.

But it is not the only force that is vectored. The term vector may refer to any directed quantity, the most elementary instance being the geometrical displacement in the previous chapter. When dealing with vectors, ordinary numbers are called *scalars*. In physics all vector quantities are ultimately related to displacements, so it is both convenient and sufficient to single them out for discussion. The displacement (x, y, z) from the origin, or from any point, has a magnitude and a direction. Attending first to the x-direction (the axes, remember, are arbitrary, but have usually been chosen to simplify some geometry), then the displacement $(1, 0, 0)$ is called the unit vector in the x-direction. It is denoted by the single bold-face symbol i. Positive or negative multiples of i can be added in any order to give a larger or smaller displacement along the x-axis. The rule is: multiplying a vector by a scalar changes its magnitude by the same factor and multiplying by -1 reverses its direction (see fig. 4.1).

Even for the simplest applications, displacements along a second, perpendicular direction are needed; thus $(0, 1, 0)$ is the unit vector along the *y*-axis and is denoted by **j**. Now, staying in the (x, y) plane, **i**- and **j**-vectors can be combined, e.g. $2\mathbf{i} - \mathbf{j} = (2, -1, 0)$ is the vector at $-26.57°$ to the *x*-axis, and of magnitude 2.236. The description is completed with the unit vector along the *z*-axis, namely $(0, 0, 1) = \mathbf{k}$.

It should be obvious from their definitions that multiples of **i, j** and **k** can be 'added' in any order to give a composite or resultant vector. Thus quite generally $a_x\mathbf{i} + a_y\mathbf{j} + a_z\mathbf{k}$ refers to a vector which has components a_x, a_y, a_z along the *x*, *y*, *z* axes respectively. The symbol **a** is used to denote such a combination. Addition of vectors **a** and **b** means adding their components or, equivalently, finding the third side of the triangle which they define (fig. 4.2). Note the possible relationship of **a** and **b** in the figures. The starting point for one is the end point of the other. Such a 'daisy-chain' relationship could be extended to add further vectors, if required.

The rules for manipulating vectors follow logically from the above definitions. First, the magnitude of **a** (usually written as *a* or $|\mathbf{a}|$) is $(a_x^2 + a_y^2 + a_z^2)^{\frac{1}{2}}$, whence it follows that $|\mathbf{a} + \mathbf{b}| \leqslant |\mathbf{a}| + |\mathbf{b}|$. The 'equals' sign holds only when **a** is parallel to **b**, and in the same sense. The vectors **a**, **b**, **c** etc., can be added or grouped in any order; $\mathbf{a} + \mathbf{b} + \mathbf{c} = \mathbf{b} + \mathbf{c} + \mathbf{a} = (\mathbf{a} + \mathbf{b}) + \mathbf{c} = (\mathbf{a} + \mathbf{c}) + \mathbf{b}$, etc.

It is useful to remember that it takes only two vectors to define a plane. The addition of any multiples of **a** and **b** defines some point in the plane which contains the triangle defined by **a** and **b**.

The combination of vectors **a**, **b**, **c** etc. may give a zero resultant, expressed as $A = 0$. This means that $A_x = A_y = A_z = 0$. Thus $\mathbf{a} = \mathbf{b}$ can be written $\mathbf{a} - \mathbf{b} = 0$, i.e. $a_x = b_x$ etc.

4.1. Vectors **i**, $-$**i** and **2i**.

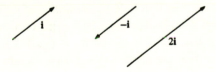

4.2. Vectors are added according to the Triangle Law. These two triangles illustrate $\mathbf{c} = \mathbf{a} + \mathbf{b}$ and $\mathbf{c} = \mathbf{b} + \mathbf{a}$ respectively

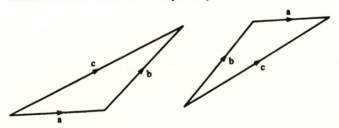

Finally, on a point of notation, the displacement from the origin, (x, y, z) defines the vector $x\mathbf{i} + y\mathbf{j} + z\mathbf{k}$, and this is commonly referred to as the *position vector* and denoted by $\mathbf{r}$.

An example from the geometry of the triangle illustrates these rules, and the power of abbreviation: fig. 4.3 shows a general triangle, *ABC*, in which the sides *BC*, *CA*, and *AB* are denoted by the vectors $\mathbf{a}$, $\mathbf{b}$ and $\mathbf{c}$ respectively. The bisectors from *A* to the mid-point *D* of *BC*, and from *B* to the mid-point *E* of *AC*, intersect within the triangle at a point *Q*. How can vectors be used to find the displacements *AQ*, *BQ*?

Starting from *B*, any point on *BE* is given by $q(\mathbf{a} + \tfrac{1}{2}\mathbf{b})$, where by varying the multiplier q from 0 to 1, every point is covered. Similarly, starting from *A* any point on *AD* (but referred to the same point *B*) is given by $-\mathbf{c} + q'(\mathbf{c} + \tfrac{1}{2}\mathbf{a})$, where q' is again required to satisfy $0 \leqslant q' \leqslant 1$. If, now, both expressions describe the same point *Q* (and both referred to *B*, remember) then

$$q(\mathbf{a} + \tfrac{1}{2}\mathbf{b}) = -\mathbf{c} + q'(\mathbf{c} + \tfrac{1}{2}\mathbf{a})$$

or

$$(q - \tfrac{1}{2}q')\mathbf{a} + \tfrac{1}{2}q\mathbf{b} + (1 - q')\mathbf{c} = 0.$$

The vector $\mathbf{c}$ can be eliminated, using $\mathbf{a} + \mathbf{b} + \mathbf{c} = 0$, to give

$$(q + \tfrac{1}{2}q' - 1)\mathbf{a} + (\tfrac{1}{2}q + q' - 1)\mathbf{b} = 0.$$

4.3. By associating the three sides of the triangle *ABC* with vectors $\mathbf{a}$, $\mathbf{b}$, $\mathbf{c}$, a problem in geometry can be solved as described in the text.

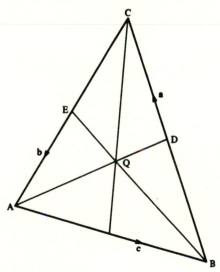

Since **a** and **b** are (in general) of finite magnitude and in different directions, this combination can only give zero if the two coefficients in brackets are both zero. A quick calculation gives $q = q' = \frac{2}{3}$, a result well known to Euclid, not to mention generations of school mathematicians.

In such geometrical applications of vectors, the notation $\overrightarrow{AB}$ is often useful in denoting the vector associated with a displacement from A to B. But note: although it is natural to draw the vector joining A to B, it is really defined only by its magnitude and direction – it does not, so to speak, belong anywhere in particular.

As a more physical example, consider the equilibrium of a particle subject to three forces $\mathbf{F}_1$, $\mathbf{F}_2$ and $\mathbf{F}_3$. Note that forces can be treated as vectors, as we implied at the beginning. We shall not justify this here. The condition that we require is just $\mathbf{F}_1 + \mathbf{F}_2 + \mathbf{F}_3 = 0$ (total force is zero). The three forces must form a triangle, as in fig. 4.4. Necessarily they are co-planar, i.e. lie in one plane.

4.4. (a) Three forces $\mathbf{F}_1$, $\mathbf{F}_2$, $\mathbf{F}_3$ which act on a particle. (b) Equilibrium condition $\mathbf{F}_1 + \mathbf{F}_2 + \mathbf{F}_3 = 0$.

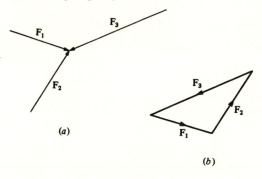

(a)

(b)

4.5. Representation of an element of area as a vector $\Delta \mathbf{S} = \mathbf{n}\Delta A$

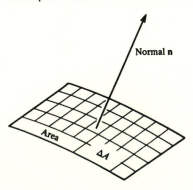

Again we must note that a vector force **F** has magnitude and direction only. Its point of application is another matter, generally requiring another vector. It is not part of the definition of **F** itself.

Other examples of vectors which crop up in physics and in other parts of this book are velocity, momentum, acceleration, angular velocity, angular momentum, and dipole moment. In problems which involve plane surfaces or small elements of area, these are also associated with vectors, with the definition *magnitude = area, direction = direction of normal*, as indicated in fig. 4.5.

Before passing on, we should note that the word *vector* is also used in a more general way, as will be explained in chapter 7.

Summary

A vector is a quantity having both magnitude and direction. Unit vectors along x, y, z axes are defined as $\mathbf{i} = (1, 0, 0)$; $\mathbf{j} = (0, 1, 0)$; $\mathbf{k} = (0, 0, 1)$. Any vector may be written $\mathbf{a} = a_x\mathbf{i} + a_y\mathbf{j} + a_z\mathbf{k}$, where the coefficients a_x, a_y, a_z may be positive or negative. The magnitude of **a** is given by $a = |\mathbf{a}| = (a_x^2 + a_y^2 + a_z^2)^{\frac{1}{2}}$.

Vectors **a**, **b**, **c** can be added of grouped in any order.

The meaning of $\mathbf{A} = 0$ is $A_x = A_y = A_z = 0$.

EXERCISES

1. A point is displaced 0.5 m down a plane inclined at 30° to the horizontal. Express this as a vector referred to axes such that
   (*i*) **j** is vertical (upwards); **k** is horizontal and perpendicular to line of greatest slope;
   (*ii*) **k** is the same and **i** is the direction of the line of greatest slope.

2. Express as a vector the body diagonal of a cube with side of unit length, referred to axes along the sides of the cube. (The body diagonal joins opposite corners.)

3. Co-planar vectors **a**, **b**, **c**, **d** satisfy $\mathbf{a} + \mathbf{b} + \mathbf{c} + \mathbf{d} = 0$. Given that $\mathbf{b} = -\mathbf{a}$ and $|\mathbf{c}| = |\mathbf{d}|$, sketch the figure formed by **a**, **b**, **c**, **d**.

4. A plane ramp is inclined at 10° to the horizontal. An uphill displacement of 2.5 m is made at an angle of 30° to the line of greatest slope. Taking axes **i**, **j**, **k** so that **k** is vertical and **j** horizontal, perpendicular to the line of greatest slope, sketch the geometry and the vectors **i**, **j**, **k**. Express the displacement as the vector $x\mathbf{i} + y\mathbf{j} + z\mathbf{k}$, and find the coefficients x, y, z.

5. A particle of speed $10^4 \, \text{ms}^{-1}$ undergoes a collision and is deflected through an angle of 35° without loss of energy. Calculate the vector

change in velocity, in terms of components referred to the original direction and a perpendicular direction in the scattering plane.

6. The vectors $\mathbf{a}$, $\mathbf{b}$, $\mathbf{c}$ form a triangle (i.e. $\mathbf{a} + \mathbf{b} + \mathbf{c} = 0$) of area A. What is the area of the triangle formed by $\lambda\mathbf{a}$, $\lambda\mathbf{b}$, $\lambda\mathbf{c}$? Give a formula which is valid for negative as well as positive λ.

7. Three forces $\mathbf{F}_1$, $\mathbf{F}_2$, $\mathbf{F}_3$ are in equilibrium. They are respectively parallel to $(10, 2, -4)$, $(5, 2, 0)$, and $(0, 1, 2)$. If $\mathbf{F}_1$ has magnitude $2\sqrt{3}$ units, what are the magnitudes of $\mathbf{F}_2$, $\mathbf{F}_3$?

8. A patrol-ship sailing steadily at $20\,$km/h in a direction $30°\,$W of N observes a yacht in the direction $40°\,$E of N at a range 1 km. Ten minutes later the yacht's position is noted as $45°\,$E of N at a range of $1.2\,$km. Assuming the yacht to be on a steady course, find its speed and direction by graphical means.

9. The three quantities μ_x, μ_y, μ_z defined in chapter 3 are the components of the (vector) dipole moment. In what direction would this point for the NH_3 molecule, assuming that the H and N atoms have net positive and negative charges, respectively?

10. Three concurrent edges of a solid cube are denoted by the unit vectors $\mathbf{i}$, $\mathbf{j}$, $\mathbf{k}$. The cube is first rotated $45°$ (anticlockwise direction) about the $\mathbf{k}$-edge. Show that the new $\mathbf{i}$, $\mathbf{j}$ edges become
$\mathbf{i}' = 2^{-\frac{1}{2}}(\mathbf{i} + \mathbf{j}), \mathbf{j}' = 2^{-\frac{1}{2}}(-\mathbf{i} + \mathbf{j})$
If the cube is now further rotated $45°$ about the $\mathbf{i}'$ direction, find the final $\mathbf{k}$ direction. If the order of the two rotations were reversed, would the final $\mathbf{k}$ direction be the same?

5
The scalar product

In mechanical drawings the orientation of the design is chosen purely for convenience. So far as the builder is concerned, it is only the relative orientation of the elements of the design that matters. Indeed, some details may be more clearly understood if the print is turned through 90°, or even read upside down.

Mathematical description of the geometrical aspect of physical phenomena is rather like this. The vector notation which uses a single symbol is handy, precisely for the reason that it is independent of reference directions. Of course at some stage in a practical calculation it usually becomes necessary to specify actual components of velocity, force etc., and these will depend upon the choice of axes. Even here, however, some geometrical features are the same whatever axes are used. Thus the distance between two points and the angle between two directions are unaffected by rotations or shifts of axes. Such fixed quantities are called *invariants*. Another important example of an invariant is the scalar product of two vectors.

In the $\mathbf{i}, \mathbf{j}, \mathbf{k}$ notation of chapter 4, a given vector (displacement, velocity, force, etc.) can be written $\mathbf{a} = a_x\mathbf{i} + a_y\mathbf{j} + a_z\mathbf{k}$. The coefficients a_x, a_y, a_z, depend upon the axes chosen, and are referred to as the components of $\mathbf{a}$, *resolved* along the directions $\mathbf{i}, \mathbf{j}, \mathbf{k}$. Indeed the vector $\mathbf{a}$ may be represented by (a_x, a_y, a_z). For a position vector, this is just the Cartesian system introduced in chapter 3. Suppose we introduce the notation $\mathbf{a} \cdot \mathbf{i}$ to mean the component of **a**, resolved in the direction **i**. Note that, in particular, **i** has zero component resolved along **j**. This feature can be represented by $\mathbf{i}\cdot\mathbf{j} = \mathbf{j}\cdot\mathbf{i} = 0$. On the other hand, resolving **i** along itself, $\mathbf{i}\cdot\mathbf{i} = 1$. Similarly for the other pairs, $\mathbf{i}\cdot\mathbf{k} = \mathbf{k}\cdot\mathbf{i} = 0$, $\mathbf{j}\cdot\mathbf{k} = \mathbf{k}\cdot\mathbf{j} = 0$, and $\mathbf{j}\cdot\mathbf{j} = \mathbf{k}\cdot\mathbf{k} = 1$.

Now these unit vector resolutions can be used to define a general 'dot product' **a·b** of any pair of vectors.

$$\mathbf{a}\cdot\mathbf{b} = (a_x\mathbf{i} + a_y\mathbf{j} + a_z\mathbf{k})\cdot(b_x\mathbf{i} + b_y\mathbf{j} + b_z\mathbf{k})$$
$$= a_xb_x + a_yb_y + a_zb_z. \tag{5.1}$$

This is more properly called the *scalar product*. Note that it is a scalar quantity. To obtain a useful alternative formula take the x-axis parallel to **a**, and the y-axis perpendicular to **a**, in the **a**, **b** plane. Then $\mathbf{a} = a\mathbf{i}$, and $\mathbf{b} = b\cos\theta\,\mathbf{i} + b\sin\theta\,\mathbf{j}$, where θ is the angle between **a** and **b**. Figure 5.1 illustrates the geometry. Then

$$\mathbf{a}\cdot\mathbf{b} = a\mathbf{i}\cdot(b\cos\theta\,\mathbf{i} + b\sin\theta\,\mathbf{j})$$
$$= ab\cos\theta. \tag{5.2}$$

But this result does not depend on the choice of axes. This *invariance* property of the scalar product will be studied in chapter 8 (exercise 4).

To see how such a product might be useful in practice, consider fig. 5.2. How much rain is collected by a bucket per unit time? This is the 'flux' of rain through the top of the bucket. The answer depends on the area and orientation of the top of the bucket and the intensity and direction of the rain. The collecting area can be represented by a vector **A** (see chapter 4), and a second vector **R** gives the direction of rainfall and (by its magnitude) the rate at which rain impinges on one square metre of surface perpendicular to **R**. A little simple geometry will give the result $AR\cos\theta$ or **A·R** for the rate of collection of the bucket.

Another example of the use of the scalar product in physics is the formula for the work W done by a constant force **F** when its point of

5.1. Choices of axes to illustrate the meaning of the scalar product.

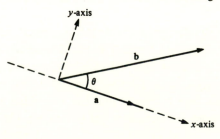

application is displaced by **x**,

$$W = \mathbf{F} \cdot \mathbf{x}$$

Note that in the special case in which **F** is perpendicular to **x**, no work is done. This applies, for example, to a particle in uniform circular motion around a point, under an attractive force directed towards it.

Three examples illustrate mathematical uses of the scalar or dot product.

Example 1: What is the angle between the vectors $\mathbf{a} = (1, 0, -5)$, and $\mathbf{b} = (3, 2, 4)$?
Answer: $\mathbf{a} \cdot \mathbf{b} = 3 + 0 - 20 = -17$; but also $\mathbf{a} \cdot \mathbf{b} = 26 \times 29 \times \cos\theta$. Equating and solving for θ, $\theta = 128.3°$.

Example 2: What is the angle between $\mathbf{a} = (1, -1, 4)$, and $\mathbf{b} = (2, -2, -1)$?
Answer: Take the scalar product, $\mathbf{a} \cdot \mathbf{b} = 2 + 2 - 4 = 0$. Thus **a** and **b** are at 90°. Quite generally, the condition that **a**, **b** are at right angles is that $\mathbf{a} \cdot \mathbf{b} = 0$.

Example 3: In chemistry, tetravalent carbon is often bonded in the so-called tetrahedral arrangement. This is illustrated in fig. 5.3 for the molecule CCl_4, carbon tetrachloride, basis of the dry-cleaning business. The three planes (drawn in perspective) are at angles 120°.

5.2. Rain passing through a planar surface (the top of a bucket).

What is sought is the tetrahedral angle between any two C–Cl bonds.

Answer: Clearly $\phi > \pi/2$, i.e. $\cos \phi < 0$. Locate the C atom at the origin O, draw $\mathbf{k}$ to point A, and $\mathbf{i}$ is normal to OA in the bond plane. Unit vector $\mathbf{j}$ (not shown) is normal to $\mathbf{i}$ and $\mathbf{k}$. Take the bond length (OA, OB, OC) as unity. Then OB is represented by the vector
$\overrightarrow{OB} = \sin \phi\, \mathbf{i} + \cos \phi\, \mathbf{k}$, and OC by $\overrightarrow{OC} = \sin \phi \cos 120°\, \mathbf{i} + \sin \phi \sin 120°\, \mathbf{j} + \cos \phi\, \mathbf{k}$. Remembering that $\cos 120° = -\frac{1}{2}$, and taking the scalar product,

$$\overrightarrow{OB}\cdot\overrightarrow{OC} = -\tfrac{1}{2}\sin^2 \phi + \cos^2 \phi.$$

But the bonds OB, OC are also at angle ϕ (this is what tetrahedral coordination means). Therefore

$$-\tfrac{1}{2}\sin^2 \phi + \cos^2 \phi = \cos \phi.$$

The required angle is a solution of the above equation. Remembering that $\sin^2 \phi = 1 - \cos^2 \phi$, the equation can be written as a

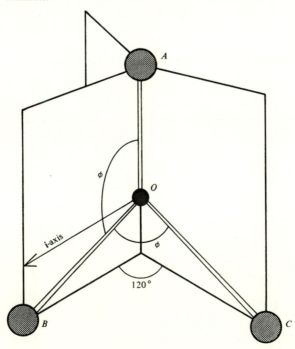

5.3. Choice of axes $\mathbf{i}$, $\mathbf{j}$ (not shown), $\mathbf{k}$ (along OA) to analyse the geometry of the CCl_4 molecule.

quadratic in $\cos\phi$,

$$\tfrac{3}{2}\cos^2\phi - \cos\phi = \tfrac{1}{2}$$

The two solutions to this quadratic equation are $\cos\phi = 1, \cos\phi = -\tfrac{1}{3}$ and clearly the latter is the required root. Thus the tetrahedral angle $\phi = 109.5°$.

In triangulation, the cosine rule, beloved of school mathematicians, can be derived using the scalar product. Suppose a vector displacement $\mathbf{a}$ is followed by a vector displacement $\mathbf{b}$, then the net displacement is $\mathbf{c} = \mathbf{a} + \mathbf{b}$, the geometry being illustrated in fig. 5.4. The magnitude of $\mathbf{c}$ is given by $|\mathbf{c}|^2 = |\mathbf{a} + \mathbf{b}|^2$. Using the fact that $c^2 = \mathbf{c}\cdot\mathbf{c}$ etc. and expanding the square, then

$$a^2 + b^2 + 2\mathbf{a}\cdot\mathbf{b} = a^2 + b^2 + 2ab\cos\theta,$$

where θ is the angle between $\mathbf{a}$ and $\mathbf{b}$. Note that θ is the external angle, as shown in the figure. If the internal angle $A\hat{C}B$ is used, then $\cos A\hat{C}B = -\cos\theta$. Of course, if $\mathbf{a}$ is normal to $\mathbf{b}$, then $\mathbf{a}\cdot\mathbf{b} = 0$ and $c^2 = a^2 + b^2$. This is not really a proof of Pythagoras' theorem since the various definitions which we have used have already assumed it.

5.4. Definition of the angle θ in the cosine rule.

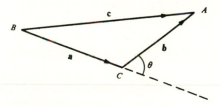

Summary

The scalar product of vectors $\mathbf{a}$ and $\mathbf{b}$ is written $\mathbf{a}\cdot\mathbf{b}$ and is given by

(i) $\mathbf{a}\cdot\mathbf{b} = a_x b_x + a_y b_y + a_z b_z$ for any set of axes x, y, z;
(ii) equivalently, $\mathbf{a}\cdot\mathbf{b} = ab\cos\theta$, where θ is the angle between $\mathbf{a}$ and $\mathbf{b}$.

The condition that $\mathbf{a}$, $\mathbf{b}$ be perpendicular is $\mathbf{a}\cdot\mathbf{b} = 0$.

EXERCISES

1. Referred to some coordinate frame, vectors $\mathbf{a}$ and $\mathbf{b}$ are given as $\mathbf{a} = (2, 1, 0)$, $\mathbf{b} = (-1, 0, 3)$. Sketch the orientations of $\mathbf{a}, \mathbf{b}$ and find the angle between them.

2. A displacement $\mathbf{a}$ of 1 m in an easterly direction is followed by a second

displacement $\mathbf{b}$ of 3 m in the direction 30° N of E. Calculate the magnitude and direction of the resultant displacement $\mathbf{c} = \mathbf{a} + \mathbf{b}$. The $\mathbf{c}$-direction is now adopted as an x-axis, and a y-axis is drawn perpendicular to it. What are the components (a_x, a_y) and (b_x, b_y) with respect to these axes?

3. The vector $\mathbf{r}$ is drawn from a fixed point (origin O) to a point P (in a fixed plane). A second vector $\mathbf{p}$ is drawn from P to another fixed point whose vector position referred to O, is $\mathbf{c}$. Given that $\mathbf{p}$ remains perpendicular to $\mathbf{r}$, then by considering the magnitude of the vector $\mathbf{r} - \frac{1}{2}\mathbf{c}$, or otherwise, show that the point P always lies on a circle. Calculate its radius and the vector position of its centre.

4. A point on the circumference of a circle is connected by straight lines to the ends of a diameter. Use vectors to show that the chords thus formed are at right angles to one another. [Hint: Denote the three radii concerned by $\mathbf{a}$, $-\mathbf{a}, \mathbf{b}$.]

5. A regular tetrahedron (each face is an equilateral triangle) has its base in the (x, y) plane and one corner at the origin. Using unit vectors $\mathbf{i}, \mathbf{j}, \mathbf{k}$ show that the three concurrent edges may be described by the vectors $\mathbf{i}, \frac{1}{2}\mathbf{i} + \frac{1}{2}\sqrt{3}\mathbf{j}$ and $\frac{1}{2}\mathbf{i} + (2\sqrt{3})^{-1}\mathbf{j} + \sqrt{2/3}\mathbf{k}$. (Hint: check angles between edges.)

6. A rectangular door of dimensions 2.2 m × 1.0 m sits in a frame of (nearly) the same dimensions. A diagonal line is drawn from the lower corner of the hinge side to the opposite upper corner. The door is now opened halfway, i.e. at an angle of 45° to the frame. Find the angle between the diagonal of the frame and the corresponding diagonal of the opened door. (Hint: Let unit vectors $\mathbf{i}, \mathbf{j}$ denote the directions of the horizontal and vertical frame edges, respectively, and let $\mathbf{k}$ be the unit vector perpendicular to $\mathbf{i}, \mathbf{j}$. Express the diagonal in terms of $\mathbf{i}, \mathbf{j}, \mathbf{k}$.)

7. A plane surface is inclined at 30° to the horizontal. A smooth groove is made in the plane at 25° to the line of greatest slope. What is the gravitational acceleration of a particle sliding freely in this groove?

8. If $\mathbf{r}$ is the position vector, show that $\mathbf{r} \cdot \mathbf{r} = x^2 + y^2 + z^2$. What surface is represented by the equation $\mathbf{r} \cdot \mathbf{r} = 1$?

9. Describe the surface represented by $\mathbf{c} \cdot \mathbf{r} = 1$, where $\mathbf{c}$ is some constant vector.

10. The potential energy associated with a dipole moment $\boldsymbol{\mu} = (\mu_x, \mu_y, \mu_z)$ in a uniform electric field $\mathbf{E}$ is given by $\phi = -\boldsymbol{\mu} \cdot \mathbf{E}$. For what direction of $\boldsymbol{\mu}$ is this (a) minimised, (b) maximised, (c) zero?

6

The vector product and rotation

The movement of a solid object by a force is practical dynamics in action. Consider for example a ship's anchor being dragged by its cable and digging into the sea-bed. How does the anchor move, and in response to what forces? There is the tension in the cable, the reaction and resistance of the sea-bed material, the weight of the anchor, and so on. (One of the authors can claim some expertise in this matter, having been involved in a court case which centred upon it.) The forces are vectors (chapter 4) – they can be resolved in any convenient direction, and they can be added according to the vector rule to give a resultant. There is a theorem of dynamics which states that the centre of mass of the anchor moves in response to this force, as if all the mass of the anchor were concentrated there. But the various forces act at different points. The cable tension acts at the end of the shank, the soil/sand reaction at the fluke-centre, and so on. Accordingly, they can also *rotate* the anchor.

What is the appropriate combination of forces with which to consider this aspect of the problem? The answer is the total *torque* or moment of the forces about some chosen point. Another theorem of dynamics prescribes the rotational motion of the body about the point, in response to the torque. In principle, the body might rotate in any direction – correspondingly, torque itself has a direction and is a vector quantity.

The reader may well be familiar with the use of torques in simple cases such as the theory of the beam balance, or the practical problem of removing the nuts when changing a wheel. In these cases, the direction of rotation is fixed and the need for vectors is not so evident. Motor mechanics use torque-wrenches, which tell them what torques they are applying, but they seldom know much about vectors. They are only concerned with the magnitude of the torque.

More general problems of dynamics require the full vector definition of the torque. Referring to fig. 6.2, one form of this is: (*a*) the magnitude of the torque τ with respect to the chosen origin is $rF \sin \theta$; (*b*) the direction of the torque is perpendicular to **F** and **r** (the 'plane of the paper'). Up or down? The rule is that **r**, **F**, τ form a 'right-handed set'. In this case, this requires that τ points into the plane of the paper, so that **r**, **F**, and τ are related like

6.1. Sketch of a modern anchor.

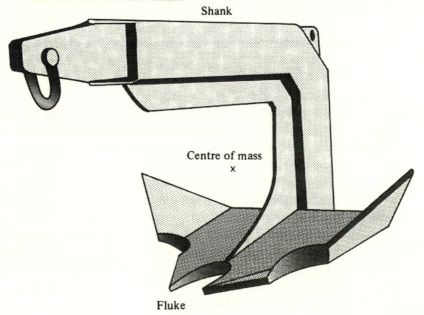

Shank

Centre of mass
x

Fluke

6.2. Two vectors **F**, and **r**, defining a force and its point of application, which are combined to define the corresponding *torque* about the origin, as explained in the text.

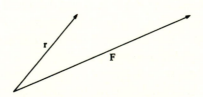

the thumb, first and second fingers (in that order) of your right hand. This is the rule which we postponed from chapter 3.

All of this can be greatly compressed by the adoption of another kind of product for vectors, so that

$$\tau = r \times F \tag{6.1}$$

where $\times$ denotes 'vector product' or 'cross product'. We can take the above statements as our definition of this, or we can resort to component notation, in which case

$$(\tau_x, \tau_y, \tau_z) = (yF_z - zF_y, zF_x - xF_z, xF_y - yF_x). \tag{6.2}$$

It is not self-evident that this is the same definition – you should try a few special cases to satisfy yourself of this. Using either definition, any two vectors can be combined to form their vector product. The word 'moment' implies the operation $r \times$. This can also be applied to momentum to give angular momentum.

If our anchor is not to rotate, the total torque (vector sum of torques due to individual forces) about the centre of mass must be zero. We shall let that particular case rest there.

6.3. Rotation of a rigid cube about a body diagonal, which joins opposite corners.

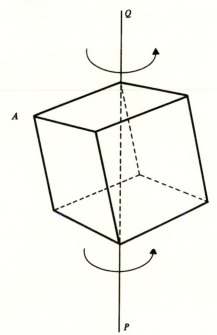

There is another related use of this kind of product, to describe rotation itself. The mathematical expression of rotation is independent of the shape of the object, but to simplify matters the anchor is exchanged for a rigid cube. Figure 6.3 shows such a cube fixed at opposite corners and free to rotate about the diagonal joining them.

It is apparent that a rotation of the body about its fixed axis PQ means that any point such as A (internal points as well) describes a circular arc. Note that whatever the position,

(*i*) each arc lies in a plane normal to the axis PQ;
(*ii*) the radius depends upon the position of the point within the body;
(*iii*) the angle of rotation is the same for all points.

To find the radius for some point may involve a bit of geometry. Thus for a corner A, fig. 6.4 shows a triangle drawn in the plane through the axis PQ and the corner. The cube side is taken as having unit length. The centre of the circle is at C and the radius is CA.

It happens that the triangle ACP is the same shape as the triangle QAP, the side AQ corresponding to the side CA. Comparing AP with QP gives the reduction factor as $\sqrt{(2/3)}$. It follows that the required radius $CA = \sqrt{(2/3)}$.

The properties (*i*), (*ii*) and (*iii*) of a rotation can be concisely expressed using vectors. A rotary displacement through some finite angle is best regarded as resulting from a succession of very small rotations, as from A to A' in fig. 6.5. For all practical purposes the arc AA' now becomes a very small linear or straight displacement.

It should be clear from this figure and from (*i*) that the short arc-displacement AA' is normal to the axis PQ. In the limit of very small $\Delta\phi$, the displacement AA' also becomes normal to the radius vector CA. Suppose now the axis direction is denoted by the unit vector **k** (i.e. the z-direction is

6.4. The corner point A is related to the axis of rotation as shown.

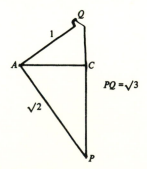

taken parallel to PQ). At the same time, the direction CA is taken as $\mathbf{i}$. Then as shown in fig. 6.6 the short displacement AA' is in the $\mathbf{j}$-direction. As a vector therefore AA' is represented by $CA\Delta\phi\mathbf{j}$, where CA here denotes the length of the line joining C to A. Now A stands for any point within or on the body. If an origin is chosen at some fixed point (so it must be on the axis PQ) the position vector of A generally may be denoted by $\mathbf{r}$. The rotation therefore shifts $\mathbf{r}$ to the new position, $\mathbf{r} + \Delta\mathbf{r}$, where $\Delta\mathbf{r} = CA\Delta\phi\mathbf{j}$.

The radius CA is just the resolution of $\mathbf{r}$ along the direction $\mathbf{i}$. In fact, referring to fig. 6.7, $\mathbf{r} = r\cos\theta\mathbf{k} + r\sin\theta\mathbf{i}$, where the angle θ is made by $\mathbf{r}$ with $\mathbf{k}$.

6.5. Rotation through a small angle $\Delta\phi$ about the axis PQ.

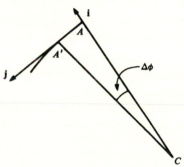

6.6. Introduction of axes $\mathbf{i}$, $\mathbf{j}$, $\mathbf{k}$ (not shown) to make a vector representation of rotation.

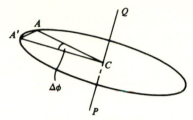

6.7. The third axis $\mathbf{k}$ and the position vector $\mathbf{r} = \overrightarrow{OA}$ are shown.

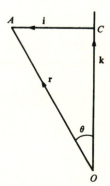

The geometry of rotation is now seen to be inescapably three-dimensional. There is the rotation axis along $\mathbf{k}$, the radius vector along $\mathbf{i}$, and the displacement vector $\Delta\mathbf{r}$ along $\mathbf{j}$. Although these relations are already adequately expressed, they can be summarised in an elegant and useful way through the cross-product notation which gives a compact expression for $\Delta\mathbf{r}$. The rotation axis vector $\mathbf{k}$ multiplied by $\Delta\phi$ provides a vector measure of the rotation. It expresses the direction and sense of the rotation axis, as well as the angle of rotation. Its magnitude is still $\Delta\phi$ of course, so that $\Delta\phi\mathbf{k}$ might be denoted by $\Delta\boldsymbol{\phi}$. Using the expression for the vector position $\mathbf{r}$ given above, $\Delta\mathbf{r}$ can very simply be written

$$\Delta\mathbf{r} = \Delta\boldsymbol{\phi} \times \mathbf{r} \tag{6.3}$$

To see this it is sufficient to note that in the cross product, $\Delta\phi\mathbf{k} \times r\cos\theta\,\mathbf{k}$ $= 0$, and $\Delta\phi\mathbf{k} \times r\sin\theta\,\mathbf{i} = \Delta\phi r\sin\theta\,\mathbf{j}$ as required. Note that

$$\mathbf{i} \times \mathbf{j} = \mathbf{k}, \mathbf{j} \times \mathbf{k} = \mathbf{i}, \mathbf{k} \times \mathbf{i} = \mathbf{j};$$
$$\mathbf{i} \times \mathbf{i} = \mathbf{j} \times \mathbf{j} = \mathbf{k} \times \mathbf{k} = 0; \mathbf{i} \times \mathbf{j} = -\mathbf{j} \times \mathbf{i}, \text{ etc.,} \tag{6.4}$$

for a right-handed set $\mathbf{i}, \mathbf{j}, \mathbf{k}$ of orthogonal unit vectors.

At this point it should be admitted that this notation is really not consistent with the notion that $\Delta\boldsymbol{\phi}$ is the increment of some vector $\boldsymbol{\phi}(t)$ as is the case with $\Delta\mathbf{r}$. Any attempt to define $\boldsymbol{\phi}(t)$ will in general lead to difficulties (see exercise 36.41). Further developments require rather more investment by way of physical and mathematical capital. To conclude, however, the general motion of a rigid body about a fixed point (rather than a fixed axis) can be described by allowing the rotation axis itself to rotate. The relevant theory will be mentioned again in chapter 27.

Finally, there are important rules to be learned for juggling with vector and scalar products. From the component definition two useful identities are easily proved:

(i) $\mathbf{a} \cdot \mathbf{b} \times \mathbf{c} = \mathbf{b} \cdot \mathbf{c} \times \mathbf{a} = \mathbf{c} \cdot \mathbf{a} \times \mathbf{b}$ $\qquad\qquad$ (6.5)

(ii) $\mathbf{a} \times (\mathbf{b} \times \mathbf{c}) = (\mathbf{a} \cdot \mathbf{c})\mathbf{b} - (\mathbf{a} \cdot \mathbf{b})\mathbf{c}$ $\qquad\qquad$ (6.6)

Note that $\mathbf{b} \times \mathbf{c}$ is a vector normal to the $\mathbf{b}, \mathbf{c}$ plane, so that $\mathbf{a} \times (\mathbf{b} \times \mathbf{c})$ must be a vector in the $\mathbf{b}, \mathbf{c}$ plane. But any vector in this plane can be expressed as a combination of $\mathbf{b}$ and $\mathbf{c}$, and it is easy to show that the correct combination is as above, e.g. by examining components.

Another useful identity is got from (ii) by writing for $\mathbf{a}$ the vector $\mathbf{b} \times \mathbf{c}$, and then rearranging the order using (i) to give

(iii) $(\mathbf{b} \times \mathbf{c}) \cdot (\mathbf{b} \times \mathbf{c}) = |\mathbf{b} \times \mathbf{c}|^2 = b^2 c^2 - (\mathbf{b} \cdot \mathbf{c})^2.$ $\qquad\qquad$ (6.7)

Summary

The vector product $\mathbf{a} \times \mathbf{b}$ is defined in two equivalent ways.

(*i*) $\mathbf{a} \times \mathbf{b}$ is the vector normal to $\mathbf{a}$ and $\mathbf{b}$ (consistent with the right-hand rule), and having magnitude $ab \sin \theta$, where θ is the angle between $\mathbf{a}$ and $\mathbf{b}$.

(*ii*) Taking an arbitrary set of axes with unit vectors $\mathbf{i}, \mathbf{j}, \mathbf{k}$,
$$\mathbf{a} \times \mathbf{b} = (a_y b_z - a_z b_y)\mathbf{i} + (a_z b_x - a_x b_z)\mathbf{j} + (a_x b_y - a_y b_x)\mathbf{k}.$$

It is used to define torque as $\mathbf{r} \times \mathbf{F}$.

If a rigid body is rotated through a small angle $\Delta\phi$ about a fixed axis, then any point $\mathbf{r}$ of the rigid body makes a short displacement $\Delta\mathbf{r}$ given by $\Delta\mathbf{r} = \Delta\boldsymbol{\phi} \times \mathbf{r}$, where $\Delta\boldsymbol{\phi}$ is in the direction of the axis, and has magnitude $\Delta\phi$.

Identities

$$\mathbf{a} \cdot \mathbf{b} \times \mathbf{c} = \mathbf{b} \cdot \mathbf{c} \times \mathbf{a} = \mathbf{c} \cdot \mathbf{a} \times \mathbf{b}$$
$$\mathbf{a} \times (\mathbf{b} \times \mathbf{c}) = (\mathbf{a} \cdot \mathbf{c})\mathbf{b} - (\mathbf{a} \cdot \mathbf{b})\mathbf{c}$$
$$(\mathbf{b} \times \mathbf{c}) \cdot (\mathbf{b} \times \mathbf{c}) = |\mathbf{b} \times \mathbf{c}|^2 = b^2 c^2 - (\mathbf{b} \cdot \mathbf{c})^2$$

EXERCISES

1. Referred to some coordinate frame (x, y, z), vectors $\mathbf{a}$ and $\mathbf{b}$ are given as $\mathbf{a} = (2, 1, 0)$, $\mathbf{b} = (-1, 0, 3)$. Calculate the components of the vector $\mathbf{c} = \mathbf{a} \times \mathbf{b}$, and find the angles made by $\mathbf{c}$ with each of the axes x, y, z.

2. A cube $1.0\,\text{m} \times 1.0\,\text{m} \times 1.0\,\text{m}$ rotates $1°$ about its body diagonal. Use the formula $\Delta\mathbf{r} = \Delta\boldsymbol{\phi} \times \mathbf{r}$ to calculate the magnitude of the displacement of each of its eight corners. Is your calculation exact?

3. A closed plane figure is formed by the vectors $\mathbf{a}, \mathbf{b}, \mathbf{c}, \mathbf{d}$ (i.e. $\mathbf{a} + \mathbf{b} + \mathbf{c} + \mathbf{d} = 0$). Describe the shape of the figure if it is known that $\mathbf{a} \times \mathbf{c} = \mathbf{b} \times \mathbf{d} = 0$.

4. Verify the formula $|\mathbf{b} \times \mathbf{c}|^2 = b^2 c^2 - (\mathbf{b} \cdot \mathbf{c})^2$ for $\mathbf{b} = x\mathbf{i} + y\mathbf{j}$, $\mathbf{c} = y\mathbf{i} - x\mathbf{j}$.

5. Evaluate $\mathbf{r} \cdot [\boldsymbol{\omega} \times (\boldsymbol{\omega} \times \mathbf{r})]$, where $\boldsymbol{\omega} = \omega_x \mathbf{i} + \omega_y \mathbf{j} + \omega_z \mathbf{k}$, and $\mathbf{r} = x\mathbf{i} + y\mathbf{j} + z\mathbf{k}$. (Hint: use a transforming identity first.)

6. A chair (with vertical legs) is rotated $45°$ clockwise about a back leg. Following this it is tilted backwards $45°$ on its two back legs. If the order of these rotations is reversed, explain why the same position is not reached.

7. The orientation of a cube is described by unit vectors $\mathbf{i}, \mathbf{j}, \mathbf{k}$ fixed in the cube and parallel to the edges. The cube is first rotated a small angle $\Delta\phi$ about $\mathbf{i}$. Use the rotation formula to show that

$$\Delta\mathbf{j} = \Delta\phi \mathbf{k}, \quad \Delta\mathbf{k} = -\Delta\phi \mathbf{j}.$$

The cube is now further rotated through the same angle $\Delta\phi$ about the new **j**-direction. Show that if $(\Delta\phi)^2$ can be neglected, then the final orientation is equivalent to a single rotation of $\Delta\phi$ about the direction $\mathbf{i} + \mathbf{j}$.

8. An electric dipole lying in the (x, y) plane consists of a charge e at (x_1, y_1), together with a charge $-e$ at (x_2, y_2). In the same plane a uniform electric field $\mathbf{E}$ is imposed, whose components are E_x, E_y. By evaluating the moments of the resulting forces (field times charge), show that the net torque can be expressed as the vector $\boldsymbol{\mu} \times \mathbf{E}$, where $\boldsymbol{\mu}$ is the dipole moment vector, namely $e(x_1 - x_2, y_1 - y_2)$.

9. If the right-hand rule proves difficult to remember, it can be augmented with a prescription like 'the right-hand set of axes is labelled in the same sense as the rotation of the thumb and fingers when opening a door with one's right hand'. Give some equivalent aide-mémoires, based on doorknobs, screwdrivers, water faucets, off-breaks in cricket or whatever.

10. When an electron moves with velocity **v** in a magnetic field **B**, it is subject to a force $-e\mathbf{v} \times \mathbf{B}$, where e is the magnitude of the charge of an electron. Draw a sketch to illustrate the relationship of **v**, **B** and this force. Explain qualitatively how this results in a spiral motion of an electron in a uniform magnetic field.

7
Matrices in physics

As explained in the previous chapters, a physical vector such as a velocity, force or torque is described by its components referred to a fixed set of axes. Three dimensions would be usual but this may be reduced to two in some cases, or even (in relativistic physics) increased to four, reckoning time as a dimension. Also, the idea of a set of variables as constituting a 'vector' can be extended to non-geometrical systems whose 'dimensions' then refer to the number of degrees of freedom of the system, that is, the number of variables needed to identify what state it is in. Two such vectors may be linearly related through an array of coefficients or matrix. Here we discuss the use and basic properties of matrices, and in the following chapter we outline their more formal manipulation.

An illustration is provided by the analysis of electrical circuits. The standard approach is to regard the circuit as consisting of loops. This requires us to introduce as many currents (as variables) as there are loops. The mathematical connection between the loop voltages and currents can be represented by a square array or *matrix* of impedance coefficients, as follows.

Consider the 2-loop circuit shown in fig. 7.1. In the left-hand loop a current I_1, driven by the battery potential V_1, flows through the series resistances R, R_1; in the right-hand loop, the current I_2 driven by V_2 flows through R and R_2. The total current flowing through R is $I_1 + I_2$. Equating the voltages V_1, V_2 to the sum of resistive falls in potential for each circuit (each given by Ohm's Law) gives the following equations:

$$V_1 = (R_1 + R)I_1 + RI_2$$
$$V_2 = RI_1 + (R_2 + R)I_2.$$

(7.1)

This pair of simultaneous equations relates the loop currents I_1, I_2 to the voltages V_1, V_2. The resistive coefficients which characterize the relation can be set out as a table:

Table 7.1 *Coefficients in the linear relation between voltages and currents.*

	I_1	I_2
V_1	$R_1 + R$	R
V_2	R	$R_2 + R$

The first row gives the coefficients of I_1, I_2 to form V_1, while the second row gives the corresponding coefficients for V_2. Such a 2×2 array of coefficients can be regarded separately from the voltages and currents which they link, and (enclosed by brackets) written as

$$\begin{bmatrix} R_1 + R & R \\ R & R_2 + R \end{bmatrix}.$$

(7.2)

This array of coefficients extends the idea of a single resistive coefficient (for one loop) to a matrix of coefficients for the double loop. The idea can be still

7.1. Electrical circuit consisting of two loops in which currents I_1 and I_2 flow, combining to give current $I_1 + I_2$ through the resistance R.

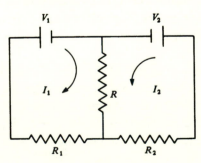

further extended to a system of many loops. The present case is described by a 2 × 2 matrix (2 rows × 2 columns), or 2-dimensional matrix.

For a particular circuit the matrix elements are fixed and serve to relate the variable voltages and currents. Suppose for example $R_1 = 3.0$ kilohms, $R_2 = 5.0$ kilohms, and $R = 1.0$ kilohm. Then the equations (7.1) can be written in matrix form as

$$\begin{bmatrix} V_1 \\ V_2 \end{bmatrix} = \begin{bmatrix} 4 & 1 \\ 1 & 6 \end{bmatrix} \begin{bmatrix} I_1 \\ I_2 \end{bmatrix} \tag{7.3}$$

where the voltages are measured in volts and the currents in milliamps. If we think of the right-hand side as a 'product' of two matrices (a single column is a 1 × 2 matrix), the 'multiplication' rule must be: to find the element belonging to a particular row and column, multiply the elements of that row of the first matrix into the elements of that column of the second, and sum. This rule is quite general. Note that the order of the matrices must not be changed.

Applying this rule to (7.3),

$$V_1 = 4I_1 + I_2, \quad V_2 = I_1 + 6I_2. \tag{7.4}$$

The pairs of variables (I_1, I_2) and (V_1, V_2) are usually described as 'vectors' though evidently they have nothing to do with the geometry of real space. In this illustration there are two components (they refer to a space of two dimensions) but in general there is no limit on the dimensionality. As written here they are described as column vectors, as distinct from row vectors, written horizontally. With such terminology two vectors can also be multiplied together to form a 'scalar product', which appears as

$$[x_1, x_2] \begin{bmatrix} y_1 \\ y_2 \end{bmatrix} = x_1 y_1 + x_2 y_2. \tag{7.5}$$

Note that this is a strict application of the matrix multiplication rule.

7.2. The rule of matrix multiplication requires that the number of columns of the first matrix equals the number of rows of the second matrix. The other dimensions determine the dimensions of the product matrix, as shown.

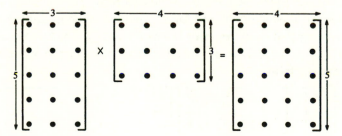

For some purposes it is necessary to express (I_1, I_2) in terms of (V_1, V_2), which just requires a solution of the simultaneous equations (7.4) (generally (7.1)) to give

$$I_1 = 0.261\, V_1 - 0.044\, V_2, \quad I_2 = -0.044\, V_1 + 0.174\, V_2, \qquad (7.6)$$

where the coefficients of V_1, V_2 are in units of kilohm^{-1}. These solutions can also be written in matrix form, namely

$$\begin{bmatrix} I_1 \\ I_2 \end{bmatrix} = \begin{bmatrix} 0.261 & -0.044 \\ -0.044 & 0.174 \end{bmatrix} \begin{bmatrix} V_1 \\ V_2 \end{bmatrix}. \qquad (7.7)$$

If we substitute back (7.6) into (7.4) we will clearly get $V_1 = V_1$ and $V_2 = V_2$. Let us see how this would proceed in matrix notation using (7.7) and (7.3) instead. These are combined according to

$$\begin{bmatrix} V_1 \\ V_2 \end{bmatrix} = \begin{bmatrix} 4 & 1 \\ 1 & 6 \end{bmatrix} \begin{bmatrix} I_1 \\ I_2 \end{bmatrix} = \begin{bmatrix} 4 & 1 \\ 1 & 6 \end{bmatrix} \begin{bmatrix} 0.261 & -0.044 \\ -0.044 & 0.174 \end{bmatrix} \begin{bmatrix} V_1 \\ V_2 \end{bmatrix}. \quad (7.8)$$

Multiplying out the product of the two square matrices, the right-hand side gives the expected result

$$\begin{bmatrix} V_1 \\ V_2 \end{bmatrix} = \begin{bmatrix} 1 & 0 \\ 0 & 1 \end{bmatrix} \begin{bmatrix} V_1 \\ V_2 \end{bmatrix}. \qquad (7.9)$$

This result assures us that matrix multiplication is the correct way of combining two successive linear combinations into one.

The square matrix in (7.9) is the 2×2 *unit matrix*. When the product of two square matrices is the unit matrix, each is said to be the *inverse* of the other.

An example of a different kind is provided by the distortion of molecules such as NH_3 or CCl_4 (fig. 3.4 and 5.3). The chemical bonds and atomic separations are not completely rigid and can be strained to produce small displacements of the atoms from their equilibrium positions. All this may be illustrated by the rather simpler carbon dioxide molecule CO_2, whose atoms lie along a straight line as shown in fig. 7.3. Consider small parallel displacements u_1, u_2, u_3 along the bond axis. (Perpendicular displacements are possible but not considered here.) The displaced atoms are no longer in equilibrium and for a given set of small displacements (u_1, u_2, u_3) are acted on by restoring forces F_1, F_2, F_3. The linear relations between F_1, F_2, F_3 and u_1, u_2, u_3 can be expressed by a matrix of dimension 3, and take the form

$$\begin{bmatrix} F_1 \\ F_2 \\ F_3 \end{bmatrix} = \begin{bmatrix} -k & k & 0 \\ k & -2k & k \\ 0 & k & -k \end{bmatrix} \begin{bmatrix} u_1 \\ u_2 \\ u_3 \end{bmatrix}, \qquad (7.10)$$

where k is a Hooke's law force constant (units Nm^{-1}). Thus if the left-hand oxygen atom is displaced by u_1 (but the other atoms are fixed, $u_2 = u_3 = 0$) then (7.10) gives the force on each atom as $F_1 = -ku_1, F_2 = ku_1$ and $F_3 = 0$. Or again, if the carbon atom is held fixed ($u_2 = 0$) but the two oxygen atoms undergo opposite displacements of equal amplitude, $u_1 = -u_3 = u$ say, then the forces become $F_1 = ku, F_2 = 0, F_3 = ku$. Generally, the matrix elements in each column provide the force constants for the corresponding displacements.

A matrix can also be used to relate the induced electric dipole moment μ of a molecule to the applied electric field E. In this case a matrix expresses the polarisability of the molecule. Only for simple molecules will it reduce to a constant. The CCl_4 molecule is one such, but CO_2 and NH_3 are obviously not – we can see different directions for E for which we would expect different values of μ. For example, we expect different results when E is parallel and perpendicular to the axis of the CO_2 molecule. In such cases, a matrix relation is necessary, although we can see that some simplification must be achievable by a sensible choice of axes. We shall return to this idea in chapter 9.

Similar remarks apply to the 'moment of inertia tensor' which relates angular momentum L and angular velocity ω for a rigid body rotating about a given point. In classical mechanics courses it comes as an unpleasant surprise that these are not parallel vectors in general and that matrix theory is needed to relate them.

Matrices may also be used for geometrical purposes, namely to express or represent mathematical effects of rotations and other operations – this is perhaps their primary use in physics. In fig. 6.4 suppose the line PQ is taken

7.3. The carbon dioxide (CO_2) molecule.

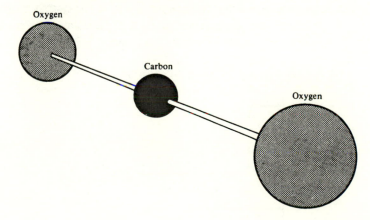

as z-axis, and perpendicular directions are added to it to form a fixed frame of axes, x, y, z. Then if the cube is rotated clockwise $90°$ about z, any point A with coordinates (x, y, z) moves to a new position A', coordinates (x', y', z').

Figure 7.4 shows that the new coordinates (x', y', z') are given by $x' = y, y' = -x, z' = z$. This change of coordinates can be represented as a matrix operation

$$\begin{bmatrix} x' \\ y' \\ z' \end{bmatrix} = \begin{bmatrix} 0 & 1 & 0 \\ -1 & 0 & 0 \\ 0 & 0 & 1 \end{bmatrix} \begin{bmatrix} x \\ y \\ z \end{bmatrix}. \tag{7.11}$$

Multiplication now refers to successive rotations. Thus a rotation through $2 \times 90° = 180°$ is represented by the product

$$\begin{bmatrix} 0 & 1 & 0 \\ -1 & 0 & 0 \\ 0 & 0 & 1 \end{bmatrix} \begin{bmatrix} 0 & 1 & 0 \\ -1 & 0 & 0 \\ 0 & 0 & 1 \end{bmatrix} = \begin{bmatrix} -1 & 0 & 0 \\ 0 & -1 & 0 \\ 0 & 0 & +1 \end{bmatrix}. \tag{7.12}$$

Summary

A matrix is an array of coefficients relating two sets of variables such as voltage and current, force and displacement. The dimension of a square matrix is the number of rows (or columns). The columns are labelled in the same order as the independent variables; the rows in the same order as the dependent variables. Matrix multiplication rule: to find the m/n element of the product matrix, multiply each element of row m in the left-hand matrix by the corresponding element in column n of the right-hand matrix, and sum, – in short, the rule is 'row-on-column'.

EXERCISES

1. A double-loop circuit (fig. 7.1) has resistance $R_1 = 1.4\,\text{k}\Omega$, $R_2 = 2.3\,\text{k}\Omega$ and $R = 0.5\,\text{k}\Omega$. Obtain the resistance matrix $\mathbf{R}$ and calculate its inverse $\mathbf{R}^{-1}$. Find the current vector for which $V_1 = 0$ and $V_2 = 2.1$ volts.

7.4. A rotation through $90°$ about the z-axis takes the point A to the point A'.

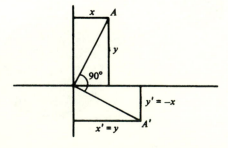

2. The quantities x_1, x_2, x_3 satisfy the equations

$$x_1 + 2.5x_2 - 0.5x_3 = u_1$$
$$2.0x_1 - x_3 = u_2$$
$$3.0x_1 - 0.3x_2 = u_3.$$

Rewrite these equations in matrix form, and find the inverse matrix.

3. Write down matrices (as described in the text) which correspond to rotations through $30°$ and $60°$ about the z-axis. Check that their product agrees with the matrix corresponding to a $90°$ rotation.

4. Work out the matrices which represent rotations which are the combinations of (a) a $30°$ rotation about the z-axis followed by a $60°$ rotation about the x-axis, (b) the same rotations in the reverse order.

5. Write down the general matrix formula for a rotation through an angle θ about the z-axis. For a small angle $\Delta\theta$, this must correspond to the formula (6.3) of the last chapter. Show that this is so, using the approximations $\sin x \simeq x$, $\cos x \simeq 1$, for small x.

6. Use the matrix equation (7.10) to show that (i) $F_1 + F_2 + F_3 = 0$, (ii) if $F_1 = F_3$ then $u_1 = u_3$. Explain why this matrix has no inverse.

7. If all possible displacements were considered, what would be the dimensions of the force matrix needed to describe (i) CO_2, (ii) NH_3, (iii) CCl_4?

8. A sample taken from the atmosphere consists of n_1 molecules of O_2, n_2 of CO, n_3 of CO_2 and n_4 molecules containing neither C or O. Express the number of C and O atoms in the sample in terms of the variables n_i using a matrix.

9. A right-angled isosceles triangular figure is formed from a deformable sheet. When the hypotenuse is stretched by a tension T_1, its length increases by an amount $u_1 = aT_1$, while the opposite vertex suffers a perpendicular inward displacement $u_2 = -cT_1$. Likewise, a tension T_2 applied perpendicular to the hypotenuse produces in this direction an extension $u_2 = bT_2$, and at the same time compresses the hypotenuse by $u_1 = -cT_2$. Express the relation between T_1, T_2 and u_1, u_2 as a matrix, and find the values of T_1, T_2 needed to extend the hypotenuse by an amount u_1 without displacing the vertex.

10. A function F is defined by the coefficients $(a_1, a_2, \ldots)$ according to

$$F = a_1 + a_2 \sin x + a_3 \cos x + a_4 x \sin x + a_5 x \cos x.$$

If F is represented by the column vector of coefficients a_i, what matrix gives dF/dx, when multiplying this column vector? Square this matrix and compare the result with the matrix which corresponds to d^2F/dx^2.

8

The transformation of matrices

Matrix notation is a good example of the way in which mathematics compresses relationships into compact forms. The price to be paid for this compression is the learning of new rules of algebra and new kinds of operation – a whole new language, in which physical scientists must be fluent.

As explained in chapter 7, the matrix extends the idea of a single coefficient R in the relation $y = Rx$ to an array of coefficients linking two sets of quantities. Such a relation is linear, meaning that if one set of quantities is scaled (i.e. each variable multiplied by the same number), then the second set is similarly scaled by the same number. A matrix is a special case of a linear operator, which has this general property.

A geometrical vector such as position is denoted by a single symbol **r**. The same notation may be used for any set of variables. Thus, a general vector **u** has elements $u_1, u_2, u_3, \cdots$.

A square array (matrix) of coefficients linearly relating **u** to a similar vector **v** is written

$$\mathbf{v} = \mathbf{Ru},\tag{8.1}$$

shorthand for

$$\begin{bmatrix} v_1 \\ v_2 \\ v_3 \end{bmatrix} = \begin{bmatrix} R_{11} & R_{12} & R_{13} \\ R_{21} & R_{22} & R_{23} \\ R_{31} & R_{32} & R_{33} \end{bmatrix} \begin{bmatrix} u_1 \\ u_2 \\ u_3 \end{bmatrix}.\tag{8.2}$$

It is helpful if vectors (column matrices) are distinguished by lower case letters in this way, but the traditions of physics sometimes overrule this convention (e.g. when the electric field vector is written as **E**).

Equation (8.2) can be written alternatively in a suffix notation. Using a single suffix to enumerate the variables (vector components), $\mathbf{u} \to u_m$, $m = 1, 2, 3$, and a double suffix to enumerate the matrix elements, $\mathbf{R} \to R_{mn}$, $m, n = 1, 2, 3$, then (8.2) becomes

$$v_l = \sum_{m=1}^{3} R_{lm} u_m. \tag{8.3}$$

Here the summation over the 'dummy' suffix m expresses the matrix multiplication rule explained in chapter 7. In general, the upper limit of the sum is simply the dimension of the vector.

For most purposes this is a clumsy notation but it often provides the neatest method of deriving basic matrix properties, being intermediate between the notations (8.1) and (8.2).

It is trivial to show that

$$\mathbf{v} = \mathbf{Au} + \mathbf{Bu} = (\mathbf{A} + \mathbf{B})\mathbf{u}, \tag{8.4}$$

whence

$$\mathbf{A} + \mathbf{B} = \mathbf{B} + \mathbf{A}. \tag{8.5}$$

Matrix multiplication has more subtle properties. It is expressed in suffix notation as

$$(AB)_{mn} = \sum_{l} A_{ml} B_{ln} \tag{8.6}$$

which is the same rule as was introduced in the last chapter. Comparing the sum with $(BA)_{mn}$ it is evident that in general

$$\mathbf{AB} \neq \mathbf{BA}, \tag{8.7}$$

that is to say, matrices in general do not commute when multiplied.

Suppose that one somewhat perversely decided to write the coefficients which make up **u** and **v** in (8.1) as row matrices instead of column matrices. How would (8.1) be rewritten? To express this neatly we need the definition of the *transpose* $\mathbf{M}^T$ of a matrix **M**. This merely exchanges rows and columns of the matrix. In suffix notation $M_{ij}^T = M_{ji}$. Thus $\mathbf{v}^T$ has one row, while **v** has one column. Our question is then answered by

$$\mathbf{v}^T = \mathbf{u}^T \mathbf{R}^T \tag{8.8}$$

as may be checked by writing out the equations in detail, or using suffix notation. The right-hand side illustrates a general rule: when transposing a product, you must reverse the order.

A matrix which equals its own transpose, $\mathbf{M}^T = \mathbf{M}$, is said to be *symmetric*. Such matrices are common in physical applications. The square matrix in eq. (7.12) is an example of a symmetric matrix. In contrast to this, the matrix in eq. (7.11) is *anti*symmetric.

There is an important aspect of linearity which has to do with changing the basis of the vectors **u** and **v**. This is best approached by an analogy with geometrical vectors. The position vector **r** can be represented as $\mathbf{r} = x\mathbf{i} + y\mathbf{j} + z\mathbf{k}$ (see chapter 4), where **i, j, k** are unit vectors taken along three perpendicular directions (defining the coordinate frame), and x, y, z are the corresponding coordinates. Now the frame, being arbitrary, may be rotated, say $\mathbf{i} \rightarrow \mathbf{i}'$, $\mathbf{j} \rightarrow \mathbf{j}'$, $\mathbf{k} \rightarrow \mathbf{k}'$ (see chapter 4 exercise 10) so that the same vector becomes $\mathbf{r} = x'\mathbf{i}' + y'\mathbf{j}' + z'\mathbf{k}'$, where x', y', z' are the new components. Quite commonly, the **i, j, k** notation is avoided and **r** written simply $\mathbf{r} = (x, y, z)$. After rotation this becomes $\mathbf{r}' = (x', y', z')$, linearly related to **r**.

Something very similar can happen to the vectors **u** and **v**; the difficulty is to grasp the idea without the visual aid of rotation, $\mathbf{i} \rightarrow \mathbf{i}'$, etc.

As explained, equation (8.1) is a linear relation between vectors **u** and **v**, typical physical instances being those of chapter 7. But the actual components of **u** and **v** depend on 'how we set up the problem', i.e. what variables we choose. Thus the components depend upon the 'frame of reference' or basis (which would be the **i, j, k** in geometrical space). Furthermore, the matrix **R** in the relation $\mathbf{v} = \mathbf{R}\mathbf{u}$ depends upon the choice of basis, and must be adjusted or transformed if the basis is changed. Such a transformation can be advantagous, that is to say, there may be a preferred basis which simplifies **R**. This is the theme of chapter 9.

In our abbreviated notation a change of basis is expressed by a relation like (8.1), namely,

$$\mathbf{u}' = \mathbf{S}\mathbf{u}. \tag{8.9}$$

It is often useful to require that the transformation be equivalent to a rotation of axes in the (abstract) space defined by the variables. We shall only consider such 'orthogonal' transformations here.

This places restrictions on the matrix elements of **S**, which we will examine in chapter 9. In this case it is obvious that the transformation is reversible, i.e. that **u** can be recovered from **u**' and written

$$\mathbf{u} = \mathbf{S}^{-1}\mathbf{u}'. \tag{8.10}$$

Here we have used the inverse matrix $\mathbf{S}^{-1}$, introduced in the last chapter, whose essential property may now be written as $SS^{-1} = S^{-1}S = 1$ (the unit matrix).

With these qualifications the transformation of the matrix **R** (note that S was introduced to transform a vector) is easily obtained. Thus 'multiplying on the right' by S, equation (8.1) becomes

$$\mathbf{v}' = \mathbf{Sv} = \mathbf{SRu}. \tag{8.11}$$

What this matrix is really doing is to combine the separate equations (8.1) with coefficients into a new set of equations. However the result is only a half-way stage since **u** also must be transformed using the inverted relation (8.10). Substituting for **u**, then

$$\mathbf{v}' = (\mathbf{SRS}^{-1})\mathbf{u}' \tag{8.12}$$

whence it follows that the transformed matrix, **R**′ is given by

$$\mathbf{R}' = \mathbf{SRS}^{-1}. \tag{8.13}$$

Since the transformation (8.13) merely changes the basis of representation of **R** it might be expected that some properties of **R** remain unaltered, just as when a pair of geometrical vectors have their frame rotated and their scalar product $\mathbf{a} \cdot \mathbf{b}$ remains the same. Quantities which do not change (i.e. are invariant) under the transformation are of particular significance. One such invariant is the sum of the diagonal elements of **R**. This is called the trace of **R**, written Tr **R**, and in suffix notation becomes

$$\text{Tr } \mathbf{R} = \sum_m R_{mm}. \tag{8.14}$$

(In continental texts the word spur is sometimes used in place of trace.)

Another such invariant is the determinant of **R**. The full theory belongs to a mathematical field much cultivated in the nineteenth century. No less a person than the Oxford creator of *Alice in Wonderland* was an early worker in this. It is said that when Queen Victoria, an admirer of *Alice*, requested a personal copy of his next book, the author sent her *The Theory of Determinants*. Presumably, she was not amused. The determinant of **R** (denoted by det **R**, or $|\mathbf{R}|$), unlike Tr **R**, depends upon each one of the elements R_{mn} in a rather complicated way. For a 2×2 matrix however, the definition is simple.

$$\det \mathbf{R} = R_{11}R_{22} - R_{12}R_{21} \tag{8.15}$$

Mathematical texts give the full definition for an $n \times n$ matrix, and its consequences, including the above property of invariance (see exercise 4).

A necessary and sufficient condition that a matrix have an inverse (enabling us to solve (8.9) for **u**) is that its determinant should not vanish.

There is in fact an explicit formula for S^{-1} involving $\det S$, which we shall not give here. However we should note that such an inverse satisfies both $SS^{-1} = 1$ and $S^{-1}S = 1$. The non-vanishing of $\det S$ is therefore the condition that a set of linear equations can be inverted to find a unique **u**. If the determinant vanishes, the solution is no longer unique.

Suppose on the other hand that the vector **u** satisfies the *homogeneous* equation,

$$\mathbf{Ru} = 0 \tag{8.16}$$

Then the same theory shows that a necessary condition for at least one non-vanishing solution for **u** is that

$$\det \mathbf{R} = 0$$

Note that the existence of an inverse would immediately imply $\mathbf{u} = 0$, since we could multiply (8.16) by $\mathbf{R}^{-1}$.

Nowadays there are generally available computer library routines to invert matrices, solve linear equations, work out determinants, etc. Matrices of dimension 10^3 or more no longer hold any terrors, when it comes to their manipulation. The task of developing an understanding of the meaning and purpose of matrix relations nevertheless remains. In this regard, the most important aspect of the theory is that covered in the next chapter.

Summary

Matrices in general do not commute when multiplied.

When a vector basis is changed by $\mathbf{u}' = \mathbf{Su}$, matrices **R** are consistently transformed as $\mathbf{SRS}^{-1}$.

The trace (sum of diagonal elements) and determinant (not fully defined here) of a matrix are examples of invariants under such a transformation.

Conditions for the existence of solutions of linear equations (see text) are expressible in terms of the vanishing of a determinant.

EXERCISES

1. Give a general proof of the stated rule that the transpose of a product of matrices is the reversed product of their transposes. [Use suffix notation.]

2. (*a*) Verify the relation

    $$\det \mathbf{AB} = \det \mathbf{A} \det \mathbf{B}$$

    for 2×2 matrices, by writing out the terms which contribute to both sides. This is in fact a useful general relation.

    (*b*) Use the above to prove that $\det \mathbf{A}^{-1} = (\det \mathbf{A})^{-1}$.

3. Using the results of exercise 2, show that the determinant of a matrix is invariant under rotations.

4. In Chapter 5, we noted that the scalar product $\mathbf{a}.\mathbf{b}$ of two vectors can be expressed in terms of their lengths and the angle between them. It is clearly invariant under rotations of axes (change of basis). Give a proof of this invariance which uses the notation of this chapter. If $M_{ij} = a_i b_j$ is regarded as a matrix, how is the scalar product related to this matrix?
   Preserving the cyclic order of i, j, k (i.e. 1, 2, 3; 2, 3, 1; 3, 1, 2), show that the elements $M_{ij} - M_{ji}$ are the components of the vector $\mathbf{a} \times \mathbf{b}$. If the components of the vectors $\mathbf{a}, \mathbf{b}, \mathbf{c}$ are regarded as the rows or columns of a 3×3 matrix $\mathbf{A}$, note that $\det \mathbf{A} = |\mathbf{a} \cdot \mathbf{b} \times \mathbf{c}|$.

5. For the deformable triangle described in chapter 7 exercise 9, the tension/displacement matrix with respect to new axes taken along the perpendicular sides is given by
$$\frac{1}{2}\begin{bmatrix} a+b+2c & a-b \\ a-b & a+b-2c \end{bmatrix}.$$
   Check the invariance of the trace and determinant under this change of axes.

6. Figure 8.1 shows a map of Scotland, divided into 11 regions. Set up a matrix $\mathbf{M}$ in which each element M_{mn} is unity if regions m and n are neighbours, zero otherwise. Find the elements in positions $(9, 9)$ and $(2, 6)$ of $\mathbf{M}^2$ and interpret your results by reference to the map. Hence find the $(3, 3)$ element of $\mathbf{M}^4$ without recourse to detailed calculations.

8.1. The regions of Scotland.

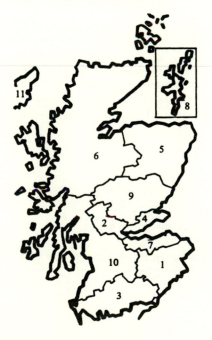

7. The matrix $S(\theta)$ is defined by
$$S(\theta) = \begin{bmatrix} \cos\theta & \sin\theta \\ -\sin\theta & \cos\theta \end{bmatrix}.$$
   (a) Show that this corresponds to a rotation through an angle θ, when used to transform vectors as in eq.(8.9).
   (b) Work out the element of $S(\theta)S(-\theta)$ and comment.
   (c) Give a formula for $[S(\theta)]^{-1}$

8. A quadratic function of two variables (x_1, x_2) can be written
$$f(x_1, x_2) = a + \mathbf{x}^T\mathbf{b} + \mathbf{x}^T\mathbf{C}\mathbf{x}$$
   where $\mathbf{x}$ is the (column) vector whose elements are x_1 and x_2, a is a constant, $\mathbf{b}$ is a constant vector and $\mathbf{C}$ is a matrix. This might describe the shape of a bowl, for example. Specify $a, \mathbf{b}, \mathbf{C}$ for the case
$$f(x_1, x_2) = 6 - 3x_1 + 4x_2 + x_1^2 + 2x_1x_2 + 5x_2^2$$
   Is your choice of $a, \mathbf{b}, \mathbf{C}$ unique?

9. Which of the following matrices represents a rotation? Discuss the significance of the other cases
$$\begin{bmatrix} 1 & 0 & 0 \\ 0 & -1 & 0 \\ 0 & 0 & -1 \end{bmatrix} \qquad \begin{bmatrix} 1 & 0 & 0 \\ 0 & -1 & 0 \\ 0 & 0 & 1 \end{bmatrix} \qquad \begin{bmatrix} -1 & 0 & 0 \\ 0 & -1 & 0 \\ 0 & 0 & -1 \end{bmatrix}$$

10. Consider the CCl_4 molecule and the axes used to describe it in chapter 5. A displacement $\mathbf{u} = (u_1, u_2, u_3)$ of the Cl atom which lies on the z-axis results in a restoring force $\mathbf{F} = (F_1, F_2, F_3)$ related to it by $\mathbf{F} = \mathbf{G}\mathbf{u}$, where the matrix $\mathbf{G}$ is given by
$$\mathbf{G} = \begin{bmatrix} a & 0 & 0 \\ 0 & a & 0 \\ 0 & 0 & b \end{bmatrix}.$$
   Here a and b are force constants associated with 'bond bending' and 'bond stretching'. Evaluate the corresponding $\mathbf{G}$ matrices for the other three Cl atoms, using the rotation matrices
$$\mathbf{R}_z\left(\frac{2\pi}{3}\right) = \begin{bmatrix} -\frac{1}{2} & \frac{\sqrt{3}}{2} & 0 \\ -\frac{\sqrt{3}}{2} & -\frac{1}{2} & 0 \\ 0 & 0 & 1 \end{bmatrix},$$
$$\mathbf{R}_y(\phi) = \begin{bmatrix} \cos\phi & 0 & \sin\phi \\ 0 & 1 & 0 \\ -\sin\phi & 0 & \cos\phi \end{bmatrix}, \quad \cos\phi = -\frac{1}{3}.$$
   Check your results by appeal to an invariance property.

9
The matrix eigenvalue equation

Of all the equations employed in mathematical physics the most widely used is the eigenvalue equation. Unlike many important equations it has not been personalised by a distinguished scientist, and indeed, would appear to have been in use long before its current name was coined. The term is hybridised, the German 'eigen' corresponding roughly to the English 'characteristic'. The eigenvalue equation appears notably in quantum mechanics, in the theory of vibrations and in celestial mechanics, where it is called the secular equation. Mathematicians employ it in the process of finding latent (or characteristic) roots of a matrix which embody most of its essential properties. But it can turn up in almost any application of matrix theory, and by extension, in the theory of differential and integral equations.

For a given square matrix $\mathbf{R}$, the eigenvector $\mathbf{u}$ and eigenvalue λ are solutions to the eigenvalue equations,

$$\mathbf{Ru} = \lambda\mathbf{u}. \tag{9.1}$$

For an eigenvector, multiplication by the matrix reduces simply to multiplication by a constant (the corresponding eigenvalue). This is reminiscent of the point which we reached in chapter 7 in discussing the polarisation of a molecule. If $\mathbf{E}$ is parallel to an eigenvector of the polarisability matrix the corresponding dipole moment is in the same direction. For the CO_2 molecule, it may be obvious that if we align $\mathbf{E}$ with the axis of the molecule, it must be so 'by symmetry'. Such arguments from symmetry are common in physical science. They have a rigorous basis in group theory, but are often consistent with simple common sense. But we do not need any symmetry or special property of a matrix to be able to find an eigenvector – every matrix has at least one!

We shall concentrate on symmetric matrices here, since they are physically significant and have especially useful properties. Note that here we use the word 'symmetric' in the special technical sense of chapter 8, which is not necessarily related to other kinds of physical symmetry, such as that which we have just discussed for CO_2.

Let us get straight to the point: *n orthogonal eigenvectors can be found for any $n \times n$ symmetric matrix.* No more, no less.

Thus for the polarisable molecule, there must be three orthogonal directions in which **E** constitutes an eigenvector and the induced electric dipole **µ** is parallel to it. Equally, any rigid body has three orthogonal directions such that angular velocity and angular momentum are parallel. These may or may not be instantly recognisable (as they are for, say, a brick) but they must exist. Returning to our arguments based on symmetry, we should note that for a cylindrically symmetric molecule such as CO_2, there are no obvious directions for the other two eigenvectors – indeed they can be any pair making up a orthogonal set in such a case. This freedom is found whenever some eigenvalues are equal and this usually is attributable to an obvious symmetry.

As a more explicit example, let us look again at the circuit problem of chapter 7. This has the advantage of familiarity and simplicity, but the vectors are more abstract, unrelated to physical space. Nevertheless, we can use the same language: 'parallel, orthogonal', etc.

Our theorem tells us that there must be two orthogonal eigenvectors $\mathbf{I}^{(1)}$ and $\mathbf{I}^{(2)}$ of the matrix (7.2). How could we set about finding them? First we write the appropriate eigenvalue equation in the form

$$(\mathbf{R} - \lambda\mathbf{1})\mathbf{I} = 0 \tag{9.2}$$

where **1** is the unit matrix (unity on the diagonal, zeros off the diagonal – not to be confused with **I**). Remember that the matrix $\mathbf{R} - \lambda\mathbf{1}$ contains an unknown quantity λ. Only for particular values of λ (eigenvalues!) will there be solutions for **I**. The condition for this was mentioned in chapter 8. It is

$$\det(\mathbf{R} - \lambda\mathbf{1}) = 0. \tag{9.3}$$

Multiplying this out for the matrix (7.2) we have

$$(R + R_1 - \lambda)(R_2 + R - \lambda) - R^2 = 0. \tag{9.4}$$

As anticipated, this will furnish *two* solutions $\lambda^{(1)}$ and $\lambda^{(2)}$, each to be substituted back into (9.2) to obtain an eigenvector. To complete this exercise without undue complexity let us now specify $R_1 = R_2 = R_0$, say, for simplicity. Then the eigenvalues are $\lambda^{(1)} = R_0 + 2R$, $\lambda^{(2)} = R_0$. Substituting back, we obtain

$$\mathbf{I}^{(1)} = 2^{-\frac{1}{2}}\begin{pmatrix} 1 \\ 1 \end{pmatrix}, \quad \mathbf{I}^{(2)} = 2^{-\frac{1}{2}}\begin{pmatrix} 1 \\ -1 \end{pmatrix}, \tag{9.5}$$

if we choose to normalise the eigenvectors (make them of magnitude unity). Note that they are orthogonal, as promised.

The first eigenvector corresponds to equal currents in the left-hand and right-hand loops (in the sense of fig. 7.1). The loop currents in R then add to give net current $2I_1$, whence it follows that $V_1 = (R_0 + 2R)I_1$. For the second eigenvector, the net current in R is zero, so $V_2 = R_0 I_2$

In many cases, equation (9.3) may have coincident roots, so that there are fewer eigenvalues than the dimension of the matrix. However, the coincident roots will still result in enough (orthogonal) solutions of (9.2) to satisfy our basic theorem. Indeed, as already remarked, there is some freedom of choice of eigenvectors in such a case. In the present case, this can be studied for the rather trivial case $R = 0$.

The orthogonal eigenvectors are a natural choice for a new basis, defining an orthogonal transformation as in chapter 8. With the new basis, the eigenvectors are just $\begin{bmatrix} 1 \\ 0 \end{bmatrix}$ and $\begin{bmatrix} 0 \\ 1 \end{bmatrix}$, and the transformed matrix is

$$\mathbf{R}' = \begin{bmatrix} \lambda^{(1)} & 0 \\ 0 & \lambda^{(2)} \end{bmatrix}. \tag{9.6}$$

This is an example of a general theorem, which is really equivalent to the one stated above. Any symmetric matrix can be diagonalised (reduced to diagonal form, with zeros off the diagonal) by an orthogonal transformation. The transformed matrix has the eigenvalues as its diagonal elements. In cases such as that of the polarisable molecule and the rotating rigid body, there are clearly great advantages in finding and using the appropriate axes.

Nevertheless, there are plenty of cases in which the physical scientist is really only interested in the eigenvalues themselves and stops short at the solution of (9.3). The dynamics of vibrations is a good example. Chemists are often concerned with the vibrational frequencies of molecules, which are directly observed by spectroscopic techniques. To model the spectrum of a given molecule, the eigenvalues of a large matrix must be found. The eigenvectors describe the vibrations appropriate to each frequency.

Until the development of modern computers, the solution of matrix eigenvalue problems for matrices of even moderate size was a forbidding one, but nowadays it is a matter of routine.

For large matrices the solution via the characteristic equation, given above, proves clumsy. Standard computer library routines, based on direct transformations of the matrix itself, are much more efficient. For symmetric matrices, their use is very straightforward, so that it requires no great expertise on the programmer's part to find the diagonalised form of, say, a $10^2 \times 10^2$ matrix. This might yield, for example, the vibration frequencies of some more complicated structure than the simple molecules considered here. An engineer might examine the vibrations of a proposed bridge in this way, by considering it to be made up of a large number of interacting components.

Summary

For a symmetric $n \times n$ matrix, n orthogonal eigenvectors can be found which satisfy the eigenvalue equation $\mathbf{Ru} = \lambda \mathbf{u}$. In cases of special symmetry, some of the eigenvalues may be equal and there is a corresponding freedom of choice for the eigenvectors. Eigenvectors and eigenvalues may be found by solving $\det (\mathbf{R} - \lambda \mathbf{1}) = 0$ for eigenvalues λ and the eigenvalues equation for $\mathbf{u}$.

EXERCISES

1. Find the eigenvalues and eigenvectors of the matrix $\mathbf{M} = \begin{bmatrix} 1 & 1 \\ 1 & 2 \end{bmatrix}$

2. Using some of the results of the exercises of chapter 8, derive the formulae

$$\det \mathbf{A} = \prod_{n=1}^{N} \lambda_n$$

$$\mathrm{Tr}\, \mathbf{A} = \sum_{n=1}^{n} \lambda_n$$

expressing the determinant and trace of a symmetric matrix of dimension N in terms of its eigenvalues λ_n. Notice that this supplies (belatedly, and for a special case) a more general definition of the determinant!

3. Multiply the vector $\mathbf{u} = \begin{bmatrix} 1 \\ 0 \end{bmatrix}$ repeatedly by the matrix $\mathbf{M}$ in exercise 1, evaluate the ratio of the components

$$\left(\mathbf{M}^n \mathbf{u} \right)_1 \Big/ \left(\mathbf{M}^{n-1} \mathbf{u} \right)_1$$

and examine its behaviour as $n \to \infty$. Comment on and suggest a generalisation of your result.

4. Our discussion of eigenvectors referred mostly to the case of a symmetric matrix. Another important special case is that of an orthogonal matrix (obeying $S^T = S^{-1}$) which represents a rotation. It always has at least one (real) eigenvector. Explain its significance. What particular rotation matrices can you think of which would have more than one eigenvector?

5. Find the eigenvalues λ_1, λ_2 of the deformable triangle tension/displacement matrix (chapter 7 exercise 9) referred to the original axes, and then to the rotated axes (chapter 8 exercise 5). Check the general results of exercise 2.

6. Find the eigenvectors of the 2×2 matrix C defined in chapter 8 exercise 8. If these were used to define new axes what would be the effect upon the original quadratic expression?

7. A laboratory demonstration of coupled oscillators consists of a pair of similar light 'dumb-bells' attached at their centres to a stiff wire stretched between two supports. At rest they are aligned parallel to each other and transverse to the wire. A small rotational displacement of one sets off a to-and-fro exchange of energy. But it is also possible to find two states of motion (normal modes) in which both dumb-bells undergo the same simple harmonic motion apart from amplitude factors θ_1, θ_2 (not necessarily positive): These must satisfy the following coupled equations
$$I\omega^2\theta_1 = \alpha(\theta_1 - \theta_2) + \beta\theta_1$$
$$I\omega^2\theta_2 = \alpha(\theta_2 - \theta_1) + \beta\theta_2$$
Here I, α, β are positive constants and ω is the angular frequency of a normal mode.

   Express the equation in terms of a matrix eigenvalue equation. Find its eigenvalues and eigenvectors. Describe the normal modes (which should accord with common sense!)

10

Exponential and logarithm functions

In seeking to express efficiently the dependence of one variable upon another, physicists have developed a repertoire of familiar functions which have well-defined properties and can be used in combination with each other to describe more complicated ones. They are the basic building blocks of practical mathematical analysis. In these two chapters we shall discuss the most elementary ones – the exponential, logarithmic and trigonometric (or circular) functions. Later, we shall see that further 'special functions' can be useful and that there is no end to this process. The skilled applied mathematician, like the skilled linguist, acquires a wide and powerful vocabulary. Usually, when we speak of an 'exact' solution to a physical problem we mean that a solution can be expressed in terms of familiar functions. Thus the availability of such a solution depends upon the range of one's vocabulary of functions.

Probably the most widely discussed, if not always identified, function in or out of physics is the exponential. Radioactive decay – 'the half-life of ^{42}K is 12.4 hours', i.e. at the end 12.4 hours the initial activity has reduced to one-half. Exponential growth – 'invest your surplus cash in X holdings at 20% (taxable)'. Sometimes threatening – 'if unchecked, the world demand for energy will double every ten years'. Let us look more closely at the everyday example of bank interest.

For a bank depositor the growth rate is often stated as the annual compound interest rate α (some fraction of the depositor's balance, per annum). Interest is commonly calculated half-yearly so that on a balance F, the interest ΔF would be $\Delta F = \alpha F \Delta t$ where $\Delta t = 0.5$. The resulting growth of F then appears as the discontinuous function shown in fig. 10.1. Here we see the regular relative (fractional) increase of a quantity F per unit time,

$$\frac{\Delta F}{F} = \alpha \, \Delta t. \tag{10.1}$$

The same equation, with negative α, can describe a decreasing trend. Time may be replaced by distance in many physical examples, as in sound attenuation through varying wall thicknesses. In scientific use (and even, it is claimed, in some banks!) the basic law (10.1) is taken to apply in the limit of small Δt. The bank computer could calculate daily or even hourly interest and the result would be to convert the discontinuous increase of fig. 10.1 to a (nearly) continuous development. It may sometimes be helpful to change the units. For example, 20% per annum could be re-expressed as $20/100 \times (365 \times 24 \times 60 \times 60)^{-1} = 6.3 \times 10^{-9}$ per second! This unit would be more appropriate to describe some cases of radioactive decay and in fact for ^{42}K could be expressed as $\alpha = -0.49\%$ per second. (For the relation of this number to the half-life of ^{42}K see exercise 5.) Introducing the symbol x (a dimensionless quantity) to denote αt (which amounts to choosing $1/\alpha$ as a new time unit), and writing $F(t) = f(\alpha t)$, the growth law 10.1 becomes

$$\Delta f = f(x + \Delta x) - f(x) = f(x)\Delta x,$$

that is,

$$f(x + \Delta x) = f(x)(1 + \Delta x). \tag{10.2}$$

10.1. Growth of a bank deposit of 100 units under compound interest paid half-yearly.

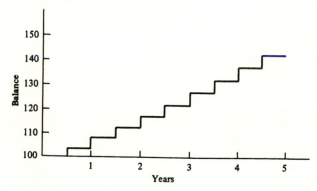

Let us take $f(0) = 1$, and (10.2) is ready for computation to see what function it defines in the limit of small Δx, when repeatedly applied.

We start from the known (or assumed) initial $(x = 0)$ value of $f(x)$ and proceed step-by-step. This process is called *iteration*. Thus, taking a reasonably small interval, say $x = 0.1$, the first step gives $f(0 + 0.1) = f(0)(0.1)$, i.e. $f(0.1) = 1.1$. Adopting the value $x = 0.1$ as the new starting point, then $f(0.1 + 0.1) = f(0.2) = f(0.1)(1 + 0.1) = 1.21$. The iteration process up to $x = 1.0$ is given in table 10.1.

The accuracy of this iteration is poor because the incremental step size $\Delta x = 0.1$ is rather too large. Remember the stipulation $\Delta t \to 0$ in (10.1). Improvement is obtained with smaller increments (and more of them!). Taking $\Delta x = 0.01$, 100 steps are needed to reach $x = 1.0$. Clearly for a general increment Δx it takes $1/\Delta x$ steps to reach $x = 1.0$. The labour of calculation is reduced by noting that according, to (10.2), each step produces the factor $1 + \Delta x$ so that $f(1.0) = (1 + \Delta x)^{1/\Delta x}$. For $\Delta x = 0.01$, $f(1.0) = (1.01)^{100} = 2.705$, somewhat larger than the previous value. There is a further appreciable but smaller increase for $\Delta x = 0.001$, which gives $f(1.0) = 2.717$. Taking $\Delta x = 10^{-6}$ (the limit of most hand calculators!), $f(1.0) = 2.7182818$. This trend continues as Δx is reduced (fig. 10.2).

The limit

$$f(1.0) = \lim_{\Delta x \to 0} (1 + \Delta x)^{1/\Delta x} \tag{10.3}$$

is denoted simply by e, which is called Napier's base. The name, in a tart remark of the distinguished Edinburgh mathematician, G. Chrystal, is 'In honour of Napier and not because he explicitly used this or indeed any other base'.

It follows from (10.2) and (10.3) that for any value of x, not just $x = 1.0$,

$$f(x) = e^x. \tag{10.4}$$

For this reason e^x is called the *exponential function* of x and often written

Table 10.1 *Iterative calculation of the function $f(x)$ [$= \exp(x)$].*

x	$f(x)$	x	$f(x)$
0	1.00	0.6	1.77
0.1	1.10	0.7	1.95
0.2	1.21	0.8	2.14
0.3	1.33	0.9	2.36
0.4	1.46	1.0	2.59
0.5	1.61		

$\exp(x)$ or simply $\exp x$. In chapter 18, we shall meet it again, as the solution of a differential equation which is just the limiting form of (10.2) as $\Delta x \rightarrow 0$.

Note the qualitative features of the exponential. As x decreases to $-\infty$, $\exp x \rightarrow 0$, while as x increases $\exp x$ increases without limit.

A basic property of $\exp x$ follows from (10.3) or (10.4), namely that for two arbitrary values x_1, x_2,

$$(\exp x_1)(\exp x_2) = \exp(x_1 + x_2). \tag{10.5}$$

Thus changing the scale of $\exp x$, i.e. multiplying by a constant C, merely shifts the function from x to $x + a$ where $C = \exp a$. Note also that $(\exp x)(\exp x) = (\exp x)^2 = \exp 2x$. In fact, in general, $(\exp x)^b = \exp bx$. Again, taking the negative of x

$$\exp(x)\exp(-x) = \exp(0) = 1, \tag{10.6}$$

so that $\exp(-x)$ is the reciprocal of $\exp x$. Figure 10.3 shows $\exp(x)$, $\exp(-x)$ and (for later reference) $\exp(-x^2)$.

If y is related to x through an exponential function, then x is related to y through a logarithmic function. Thus, if $y = \exp x$, then equivalently, $x = $ (Naperian) logarithm of y, written $x = \ln y$.

The basic property of $\ln y$ follows from (10.5), namely

$$\ln(y_1 y_2) = \ln y_1 + \ln y_2. \tag{10.7}$$

The graph of $\ln x$ plotted against x is shown in fig. 10.4. This graph is obtained from fig. 10.3 by exchanging horizontal and vertical axes.

10.2. Evaluation of $(1 + \Delta x)^{1/\Delta x}$. The dotted line denotes the limit as $\Delta x \rightarrow 0$.

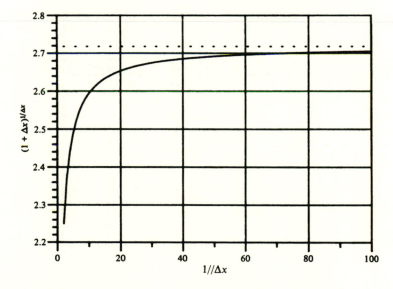

10.3. The functions $\exp x$ (full line), $\exp(-x)$ (dotted), $\exp(-x^2)$ (dashed).

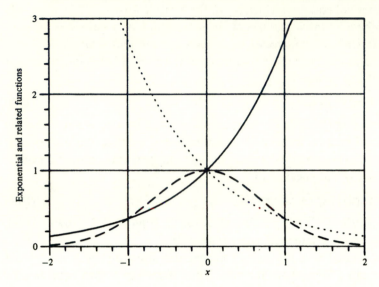

10.4. The logarithmic function $\ln x$.

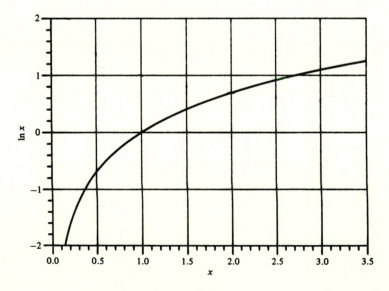

Equation (10.7) is the basis for the popularity of logarithms (other than with school pupils) since their invention by Baron Napier of Merchiston in 1614 as a means of avoiding multiplication. Napier's castle still stands in Edinburgh. While not high on the list of tourist attractions, its incorporation in a College of Science and Technology is a pleasing visual metaphor for the modern role of his function.

Often, when we meet a function in the form a^x we will find it desirable to rewrite it according to

$$a^x = e^{(\ln a)x}. \tag{10.8}$$

This is convenient for some purposes, such as differentiation (chapter 13). The equation is also useful in converting to 'logarithms with base 10', defined by

$$y = 10^{\log_{10} y}. \tag{10.9}$$

Some physical properties are traditionally quantified in terms of their logarithms. When a magnitude can vary by large factors, as in acoustic measurements, its relative change can be conveniently expressed as a power of 10. Thus a fractional increase of 3,400 could be written as $10^{3.53}$. This would be described as an increase of 35.3 decibels (db). The statement that the relative sound level at some point is 3.5 db means that the relative increase is 10 raised to the power of 3.5/10. Thus a fractional increase f corresponds to $10 \log_{10} f$ db. In electrical power usage, a different definition is used, namely, number of (power) db is given as $20 \log_{10}$ (power ratio).

Summary

Many kinds of simple growth/decay processes are described by the exponential function. It is denoted by $\exp x$ or e^x.

Basic property: $\dfrac{\exp(x + \Delta x) - \exp(x)}{\Delta x} \to \exp(x),$

as $\Delta x \to 0$ for any x.

Naperian $e = \exp(1.0)$ (Useful rule; $e^3 \approx 20$)

Basic properties: $\exp(x_1)\exp(x_2) = \exp(x_1 + x_2)$,

$\exp(-x) = 1/\exp(x)$

The Naperian logarithm $\ln x$ is the inverse function of the exponential, that is, if $x = \exp y$, then $\ln x = y$.

If a is an arbitrary positive number, and $a^x = y$ then by definition $x = \log_a y$ (logarithm to the base a) and is related to the Naperian logarithm by $\log_a y = (\ln a)^{-1} \ln y$. Thus, $\log_{10} y = 4.34294 \ln y$.

EXERCISES

1. Use the provided graph of $\exp x$ (fig. 10.3) to verify that $\exp(x_1)\exp(x_2)$ $= \exp(x_1 + x_2)$ for the value $x_1 = 0.4, x_2 = 0.6$.

2. Use figs. 10.3 or 10.4 to obtain approximate graphical solutions to the transcendental equation $\exp x = 4.5\,x$.

3. On a separate sheet of paper plot the function $2.5 \exp x$. Slide this sheet over the graph of $\exp x$ (drawn on the same scale), keeping the same x-axis and parallel y-axis, to get a coincidence of graphs. Measure the slide displacement and explain its relationship to the factor 2.5.

4. The hyperbolic trigonometric functions are defined as follows: $\cosh x$ $= \frac{1}{2}(\exp x + \exp(-x))$; $\sinh x = \frac{1}{2}(\exp x - \exp(-x))$. Prove that $(\cosh x)^2$ $- (\sinh x)^2 = 1$, for all values of x. Draw sketches of the two functions.

5. Memorise the result $2^{10} = 1024$. What percentage error is incurred by replacing 1024 by 10^3? Deduce an approximate value for $\log_2 10$ and estimate the error. Use the factorisation $2 \times 7^2 = 98$ to find $\log_{10} 7$ approximately.

6. In exponential decay, the function $A \exp(-x)$ decreases by one half for the value x given by $\exp(-x) = 0.5$, regardless of the value of A. Show that x $= \ln 2 = 0.693$.

7. The half-life of ^{42}K is 12.4 hours. Calculate the relative radioactive strength after 24.8 hours.

8. Atmospheric pressure falls off exponentially. The fractional decrease per unit height increment is $0.125\,\mathrm{km}^{-1}$. Taking the ground pressure as 1000 mbars, calculate the pressure at a height of 10 km.

9. A beam neutron of initial energy 1 Mev passes through an absorbing material and loses energy at the rate of 6% per mm. Express as an exponential function of x the neutron energy (in eV) after travelling a distance x (in mm). What thickness of absorbing material is needed to reduce the energy to 10 meV?

10. The current-voltage characteristic of a semiconductor diode is often well described by
$$I = I_0(\exp AV - 1)$$
where I_0 and A are constants. Sketch the function $I(V)$. If $A = 0.5$ volts^{-1}, what is the ratio between the currents passed by the diode under forward bias (V positive) and reverse bias (V negative) for $|V| = 2$ volts?

11

Sine and cosine functions

In the practical world of navigation and surveying, angles are measured by dividing a circle into 360 equal segments or degrees. Some engineers, out of enthusiasm for metrification, prefer to divide the quadrant into 100 grades. Either way provides a useful way of measuring angles, but for theoretical purposes, the ratio of arc length to radius provides a more natural measure (in *radians*). For the complete circle this takes the value 2π, where π is a celebrated irrational number which has been a focus of interest since the dawn of mathematics. In 1873 a Mr Shanks calculated its value to 707 decimal places. This is sufficient for most of us but mathematicians interested in the arithmetic of irrational numbers can always get more digits from the computer. The value 3.141 59 will be good enough for most purposes in this book.

So far as the geometrical use of the functions sine, cosine, etc., goes, it does not matter much whether the angle is measured in radians or degrees. From a practical point of view the radian angle of 0.785 ($=\pi/4$) is more easily recognised as 45°, but in modern physics radians are generally used.

Of the hundreds of identities that used to comprise the old science of trigonometry, one or two are worth committing to memory. The basic cosine combination formula is given by

$$\cos(A + B) = \cos A \cos B - \sin A \sin B. \tag{11.1}$$

This is most readily seen using the scalar product rule. In the (x, y) plane take vectors **a**, **b** having unit magnitude and making angles A, B respectively with the x-axis. Then resolving, $\mathbf{a} = \cos A\mathbf{i} + \sin A\mathbf{j}$, $\mathbf{b} = \cos B\mathbf{i} + \sin B\mathbf{j}$. Their scalar product $\mathbf{a \cdot b}$ becomes (see summary, chapter 5) $\cos A \cos B + \sin A \sin B$. But since the angle between **a** and **b** is $A - B$, the scalar product is also $\cos(A - B)$. The identity holds generally for all A, B. Thus, changing the sign of angle B leaves $\cos B$ unchanged but replaces $\sin B$ by $-\sin B$ to give the above quoted rule.

Again, replacing A by its complement $90° - A$ gives $\cos(90° - A + B) = \sin(A - B)$, while $\cos A \rightarrow \sin A$ and $\sin A \rightarrow \cos A$. Thus $\sin(A - B) = \sin A \cos B - \cos A \sin B$. To double the angle, put $B = A$ in the cosine sum to give $\cos 2A = \cos^2 A - \sin^2 B$. Similarly, from the sine rule, $\sin 2A = 2 \sin A \cos B$. Dividing this result by the previous one and rearranging gives $\tan 2A = 2 \tan A(1 - \tan^2 A)^{-1}$.

There is no end to the ingenious manipulation of these functions and it used to provide abundant material for stimulating and testing the young mathematical imagination. In the end however, their practical geometrical use is to obtain numerical answers to problems involving triangulation. For this purpose it is necessary to tabulate the values of the cosine and sine functions within the quadrant $0° \leqslant \theta \leqslant 90°$. It is easy enough on a digital calculator to enter the required angle, press the 'cos' button and read off the answer to eight or ten digits. But from a functional point of view, what mathematical processes define $\cos x$, $\sin x$, etc.? They have certainly not been obtained by drawing triangles and measuring ratios!

The simplest, though not the most efficient, way is to use an iterative procedure similar to that of chapter 10. It is necessary then to relate values of the cosine and sine at some value x (using radian measure for convenience) to those at the next step value $x + \Delta x$. This can be done using the addition formula (11.1), reading x for A and Δx for B, together with an important geometrical notion. As described in chapter 6 a small arc (angle Δx) becomes in the limit $\Delta x \rightarrow 0$ identical with the straight section which connects the unit radii (almost) at right angles. Thus in fig. 6.6 (taking AC as unity) the arc length $AA' = \Delta \phi$ is very nearly equal to $\sin \Delta \phi$.

Properly stated,

$$\lim_{\Delta x \to 0} \left(\frac{\sin \Delta x}{\Delta x} \right) = 1 \tag{11.2}$$

or

$$\sin \Delta x \sim \Delta x \text{ in that limit.}$$

Of course this is easily verified on the calculator but it is established here from a geometrical point of view. It will be shown in the next chapter that the error in writing $\sin \Delta x \simeq \Delta x$ is proportion to $(\Delta x)^3$.

The addition formula for cosine and sine can now be simplified by replacing $\sin \Delta x$ by Δx and $\cos \Delta x$ by 1 (because $\sin^2 \Delta x + \cos^2 \Delta x = 1$ there is a small error proportional to $(\Delta x)^2$). If Δx is sufficiently small the terms proportional to $(\Delta x)^2$ can be neglected to give the addition formulae

$$\cos(x + \Delta x) = \cos x - \Delta x \sin x$$
$$\sin(x + \Delta x) = \sin x + \Delta x \cos x$$

(11.3)

and the two functions are now ready for calculation. The procedure resembles the exponential calculation except that a double column is required. As in that instance the step size determines the accuracy – the smaller the better – but of course with more labour. Starting from $x = 0$, where $\cos x = 1$ and $\sin x = 0$ (evident from geometry), a manageable value is $\Delta x = 0.025$.

Suppose we wish to use this calculation to make an estimate of π.

It is known (by considering an equilateral triangle) that $\sin 30° = \frac{1}{2}$, so that if the calculation is continued until $\sin x$ reaches 0.5, the value of x will correspond to $\frac{1}{12}$ of the circumference of a circle. The tabulation is set out

Table 11.1 *Iterative calculation of values of sin x and cos x.*

x	$\cos x$	$\sin x$
0.0	1.0	0.0
0.025	1.0	0.025
0.050	0.999	0.050
0.075	0.998	0.075
0.10	0.996	0.100
0.125	0.994	0.125
0.150	0.991	0.150
0.175	0.987	0.175
0.200	0.983	0.200
0.225	0.978	0.225
0.250	0.972	0.249
0.275	0.966	0.273
0.300	0.959	0.297
0.325	0.952	0.321
0.350	0.944	0.345
0.375	0.935	0.369
0.400	0.926	0.392
0.425	0.916	0.415
0.450	0.906	0.438
0.475	0.895	0.460
0.500	0.883	0.482
0.525	0.871	0.504

above (table 11.1). Each entry (from the second onwards) is calculated from the line above using the iteration formulae (11.3).

The separation of the last two sine entries is 0.022 so that the sought value $\sin x = 0.5$ lies between them in the ratio $0.018/0.022 = 0.82$. The corresponding x value is therefore $0.500 + (0.82 \times 0.025) = 0.520$. This value of x (in radians) corresponds to $\frac{1}{12}$ of the perimeter. Therefore this calculation would give the semi-perimeter as $6 \times 0.521 = 3.13$, to be compared with the more precise value of π given above. The value of the present calculation is not its accuracy but the method used. It may be recalled that to get a correct 5-figure value for Napier's e, one million steps were needed. The same effort applied here would produce similar accuracy. For most practical purposes the value 3.141 59 is sufficient.

Having evaluated $\cos x$, $\sin x$, for $0 \leqslant x \leqslant \pi/6$, the doubling formulae can be used to extend the range to $\pi/3$. But since $\cos x = \sin(\pi/2 - x)$, the remainder of the quadrant is already available. To complete the remaining quadrants the addition formulae are applied to give $\sin(x + \pi/2) = \cos x$, $\sin(x + \pi) = -\sin x$, $\sin(x + 3\pi/2) = -\cos x$, etc.

A key property of the circular functions is their *periodicity*. For example $\cos(x + 2\pi) = \cos x$ so that adding any integral multiple of 2π to x leaves the cosine unchanged. Also, $\cos x$ is described as an *even* function, meaning that $\cos(-x) = \cos x$, while $\sin x$ is an *odd* function, $\sin(-x) = -\sin x$. All of

11.1. Trigonometric functions: $\sin x$ (full line); $\cos x$ (dashed); $\tan x$ (dotted).

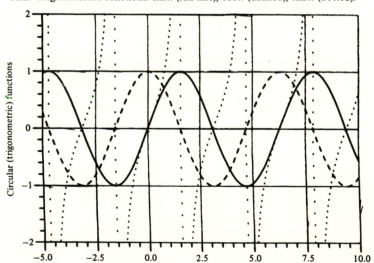

these features are displayed in fig. 11.1, which plots cos x, sin x, and tan x over a wide range.

The functions cosine and sine find their most practical use in the description of alternating current (AC) circuits. Further consideration of the theory will appear later (chapters 20, 21, 23) but some useful comments can be made even at this stage. In AC circuitry the (angular or circular) frequency is expressed in radians per second, symbol ω, where $\omega = 2\pi$ × number of cycles per second. Voltage and current response are expressed in sinusoidal form, e.g. $V(t) = V_0 \cos \omega t$, where V_0 is the peak voltage. The current response to such a driving voltage would also be sinusoidal but in general shifted in phase, and written

$$I_0 \cos(\omega t - \phi).$$

The symbol ϕ now denotes a *phase angle*, and as written, is also described as a phase lag. Note that we could expand out the cosine and consider the current to consist of two sinusoidal contributions, according to $\cos(\omega t - \phi) = \cos \phi \cos \omega t + \sin \phi \sin \omega t$. The first is 'in phase with' the driving voltage, the second is 'out of phase'.

Summary
Addition formulae for the cosine and sine

$$\cos(A + B) = \cos A \cos B - \sin A \sin B$$
$$\sin(A + B) = \sin A \cos B + \cos A \sin B$$

Doubling formulae

$$\cos 2A = \cos^2 A - \sin^2 A$$
$$\sin 2A = 2 \sin A \cos A$$

Periodicity

$$\cos(x + 2m\pi) = \cos x, m = 0, 1, 2 \text{ etc.}$$

Symmetry

$$\cos(- x) = \cos x, \sin(- x) = - \sin x$$

Useful (lowest order) approximations

$$\cos x \simeq 1, \sin x \simeq x, \quad \text{for} \quad x \ll 1$$

EXERCISES

1. Use the sine doubling formula and the identity $\cos^2\theta + \sin^2\theta = 1$ to show that $\cos 15°$ is given by $\left(\dfrac{1}{2} + \dfrac{\sqrt{3}}{4}\right)^{1/2}$. Evaluate this and check it against the tabulated cosine value.

2. (a) Use the familiar values $\sin(\pi/4) = 0.7071$, $\sin(\pi/6) = 0.5$ to obtain the value $\sin(\pi/12) = 0.2588$ by expressing it as $\sin(\pi/4 - \pi/6)$.

   (b) Use the iterative method given in the text to evaluate $\cos x$, $\sin x$ for a range of values of x starting from $\sin x = 0.2588$ and going as far as $\sin x = 0.5$. Take the incremental step as $\Delta x = 0.005$ and work to an accuracy of 0.1%. By evaluating the difference $\sin^{-1} 0.5 - \sin^{-1} 0.259$, estimate the value of π.

3. Use appropriate formulae from the summary to find the mean values of: $\sin x$, $\cos\frac{1}{2}x$, $\sin^2 2x$.

4. Find three solutions to the equation $\sin x + 0.5\cos x = 1.05$ (expressing the answers in degrees).

5. In three dimensions the coordinates (x, y, z) of a point P refer to rectangular axes OX, OY, OZ. The plane containing OZ and OP makes an angle θ with OZ, and the line OP has length r and makes an angle ϕ with OZ. Draw a perspective figure to illustrate this geometry, and hence show that
$$x = r\sin\theta\cos\phi, \quad y = r\sin\theta\sin\phi, \quad z = r\cos\theta.$$
[These are the *polar coordinates* of P]

6. Find the length in kilometres of one degree of longitude at the Earth's equator. (Take the Earth's radius as $6378\,\text{km}$.) What would be the corresponding length at a latitude of $50°$?

7. In the theory of diffraction gratings the following formula appears:
$$A = \frac{\sin Nx}{x}, \quad x = \frac{2\pi d}{\lambda}\sin\theta.$$
Taking the values $\theta = 0.10°$, $d = 10^{-5}\,\text{m}$, $\lambda = 5 \times 10^{-7}\,\text{m}$, calculate A for $N = 100$. If there is a possible error of 0.1% in d, what is the biggest useful value of N if A is required to within 15%?

8. Plot the function $\cos t - 3\sin t$ for $0 \leqslant t \leqslant \pi$. (Take intervals of 0.2.) For the resultant function determine (i) the amplitude, (ii) the phase, relative to $\cos t$.

9. The x and y coordinates of a point are given as functions of time t by
$$x = \cos\omega t$$
$$y = \sin\omega' t$$
Sketch the motion of the point
   (i) for $\omega = \omega'$

(*ii*) for $\omega = 2\omega'$

(*iii*) for $\omega = \frac{3}{2}\omega'$

10. A piece of dry ice is hung on a spring. Because the mass decreases (due to sublimation) the angular frequency of oscillation ω is a decreasing function of time. Sketch the motion (displacement versus time). Even if simplifying assumptions are made (no drag etc.) this proves a very tricky problem to solve properly, as regards the amplitude of the motion, so you should not worry too much about that aspect!

12
Graph plotting and curve sketching

An invaluable accessory to a computer or experimental system is a curve-plotter. The stony columns of data in the digital print-out come to life as plateaux, peaks, dips, shoulders and so on. Indeed, the pictorial language of numerical trends is in every sense, graphic. What business is complete without its sales chart? Nevertheless the presentation of experimental data in the form of a graph is a surprisingly modern convention. It did not become popular until the late nineteenth century. Before that, the general form of the results of experiments was described only roughly, in words. For example, Michael Faraday, who is credited (among many other things) with first recognising the characteristic property of semiconductors in 1833 simply said, 'The conducting power increases with the heat'. At that time, whole textbooks were written on 'Natural Philosophy' without graphs, even though Descartes had long since provided the conceptual framework for them. Today, such a statement would rarely be made without an accompanying graph of the data or sketch of the inferred functional form.

A graph can present experimental data or a theoretical prediction or the comparison of both in a clear and compact way. To maximise its effectiveness in your own work, you should give a little thought to the choice of origin, scales and functional forms which you use.

First, there is no necessity whatever that the meeting of the horizontal and vertical axes should coincide with the zero of either variable. For example, in plotting the current/frequency relation for an electrical circuit, only the frequencies within a factor of ten or so of the resonant frequency may be of interest. The origin can be taken at any value below this region.

In plotting mathematical functions, the origin may well be placed at the zeros of both axes. In that case it must be decided if both positive and negative sections are required. This depends on whether any symmetry is present, e.g. whether $f(-x) = (+\text{ or } -)f(x)$. The point is taken up later.

In converting numerical data or functional relations to graphical form, the next practical consideration must be that of scale. If accuracy is at all important, you should try to make your graph fill as much as possible of whatever convenient size of graph paper is available. It is wise to choose a sensible, easily managed scale rather than one which uses some awkward ratio of the intervals on the graph paper and the physical units.

While the general aim of function-plotting is to display the shape and features of a mathematical relation, the experimentalist plots graphs for a further reason, namely to extract physical parameters from a mass of data in a simple and transparent way. Consider the following example. A chemical reaction rate K (rate of composition or decomposition) may be increased by raising the temperature T according to a law $K = A \exp(-E/RT)$ (fig. 12.1) where A and E are parameters which depend upon the reacting species. (R is the universal thermal constant that turns up in the gas law $PV = RT$). By measuring the reaction rates, values of K are obtained for a range of temperature. What are now sought are the parameters A and E which characterise the reaction. Instead of plotting K against T, the

12.1. Plot of a chemical reaction rate against temperature.

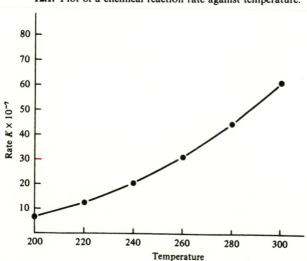

12.2. (a) Example of an Arrhenius plot on ordinary (linear) graph paper. (b) The same plot using semi-logarithmic graph paper.

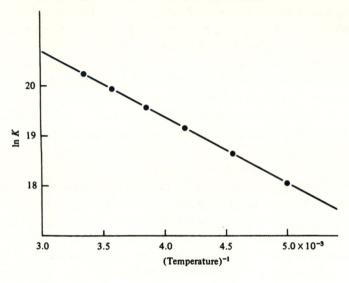

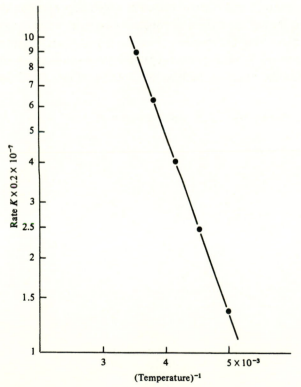

chemists make what is called an Arrhenius plot. Taking Naperian logarithms the basic law is re-expressed as $\ln K = \ln A - (E/R)T^{-1}$. Now plot $\ln K$ against T^{-1} as in fig. 12.2(*a*) and the relation becomes linear, a simple straight line with a vertical intercept of $\ln A$ and a negative slope, $-E/R$. The parameters are thus found very precisely. The same kind of plot is useful in analysing the temperature dependence of the conductivity of semiconductors which was mentioned above. This also depends exponentially on T^{-1} in many cases. It is a good general rule (although it should not be inflexible) that the variables should be chosen to make functional relationships *linear* wherever this is convenient, in analysing data. In some cases, this has become such a matter of routine that special graph paper has been developed, the use of which obviates the taking of logarithms etc.! Figure 12.2(*b*) shows the use of 'semi-log' paper to make the same plot as in fig. 12.2(*a*).

Finally, note that in plotting experimental data, it is often important to include error bars, which indicate the limits of uncertainty of the data (see chapters 1, 2). In the experiment depicted in fig. 12.3, a Coke can was rolled down an inclined plane in order to estimate g, the acceleration due to gravity. The time t taken to roll a distance l of 34 inches was measured with a sequence of values for the vertical drop h. According to theory, the value of g may be related to ht^2, if the moment of inertia of the can is known. The experiment fails completely with a full can, because of systematic errors due to swirling of the liquid. Even a freshly emptied can may contain enough

12.3. Can rolling down a ramp in a simple experiment to measure the acceleration due to gravity, g.

residual liquid to give a large systematic error! With a dry can, results shown in fig. 12.4 were obtained. Note that vertical error bars are necessary because t and h were measured rather roughly. (In some cases, horizontal error bars are also needed.) The length of the error bars is not constant. Time uncertainty dominates at large h, height uncertainty at low h – the error bars shown combine both. (For further discussion see exercise 12.2. We are indebted to Bob Walraven for this entertaining example.)

Whereas graph-plotting is concerned with the numerical representation of physical data (experimental readings or computations) curve-sketching aims to show qualitative behaviour, with a minimum of calculation. For physicists, ever being presented with new or apparently new effects, it is

12.4. Data from Coke can experiment with appropriate error bars.

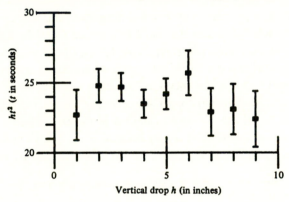

12.5. Parabolic functions, $y = f(x)$

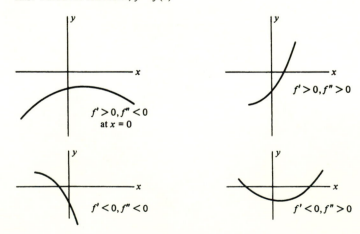

important to put a 'shape' to an unfamiliar formula. It is a useful skill, which regrettably is not always encouraged amongst tyro mathematicians.

The simplest kind of algebraic function is the polynomial, that is, a sum of powers, for example $y = c_0 + c_1 x + c_2 x^2$ where c_0, c_1, c_2 are constant coefficients of either sign. Without specifying these numerically, it is possible to sketch the general behaviour of the function.

First of all, putting $x = 0$ gives $y = c_0$, which immediately identifies c_0 as the y-axis intercept. Now increase x to a small value such that $|c_2 x^2| \ll |c_1 x|$, i.e. $|x| \ll |c_1/c_2|$. Then $y \approx c_0 + c_1 x$ so that c_1 is the initial (i.e. at $x = 0$) slope. As x increases towards $2c_1/c_2$ the $c_2 x^2$ term becomes comparable to $c_1 x$ and the curve may cross the x-axis or turn away from it, depending on the sign combinations of c_1, c_2. Assuming for definiteness that c_0 is finite and negative, and that $c_1 \neq 0$, $c_2 \neq 0$, four possibilities are shown schematically in fig. 12.5. Each of these curves is described as parabolic.

Exceptionally it may happen that $c_1 = 0$, that is, the function is flat at $x = 0$ as shown in fig. 12.6. Such flat behaviour at $x = 0$ is described as stationary – the function near $x = 0$ is neither increasing or decreasing. A positive coefficient c_2 describes an upward curvature while negative c_2 describes downward curvature. The reciprocal $1/|c_2|$ is simply the radius of the circle which exactly matches the function at $x = 0$. Note that $y = c_0 + c_2 x^2$ is an even function, that is to say, y has the same value for x and $-x$. The left-hand and right-hand sides of the sketches in fig. 12.6 are mirror images of each other.

There is still one uncovered possibility: c_1 could be finite while $c_2 = 0$. In that case it is necessary to extend the approximation to include the next non-vanishing term $c_3 x^3$. Such a point is termed a point of inflexion. Its schematic appearance is shown in fig. 12.7.

We have discussed these various cases in detail because they describe

12.6. Stationary behaviour at $x = 0$

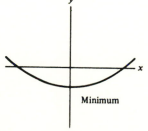

Minimum

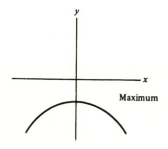

Maximum

very generally the local behaviour of more complicated functions. Close to any given point x, functions can usually be approximated by a sum of powers of x, as we shall discuss further in chapter 14.

Whenever this expansion is valid, the function is regarded as 'well-behaved' at the point in question. This is not moral approval but is used to exclude so-called 'singular' points. The simplest kind of singularity is that of the function $(x - a)^{-1}$ which diverges and switches sign as x is taken through the singular point a. Its appearance is shown in fig. 12.8. Genuine physical instances of such behaviour are hard to come by. According to the ideal gas law, $PV = RT$, with temperature T fixed the pressure P required to compress the gas to a small volume V increases according to $P = RT/V$. But this is only the right-hand side of a mathematical singularity since negative V has no meaning. A better instance occurs in the theory of resonance where a system, such as one of the molecules of chapter 7, is subjected to a force which oscillates with variable frequency. The amount of vibration that results varies as in fig. 12.8, around one of the normal mode frequencies,

12.7. Points of inflexion at $x = 0$

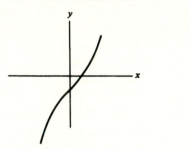
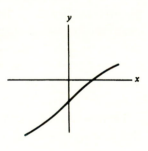

12.8. Example of a singularity at $x = 0$.

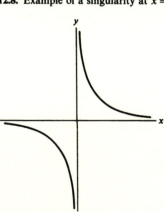

in the simplest theories. (In better ones the singularity is replaced by a pronounced wiggle.) Lastly, the gravitational potential energy at a point mass shows this kind of singularity and again it is a rather artificial (but important) theoretical construct.

This chapter can be concluded with some practical rules and tips for curve sketching. How should one set about sketching the function $y = (x + 2)^2/(x - 1)$?

(i) Look for singularities.

Clearly $x = 1$ is a simple singularity, so begin the sketch about this point.

In the vicinity of $x = 1$ the factor $(x + 2)^2$ is comparatively constant and equal to $(1 + 2)^2 = 9$, but calculation is not intended or necessary.

(ii) Look for zeros (where the function vanishes).

The factor $(x + 2)^2$ is even about $x = -2$, i.e. does not change sign as x

12.9. Curve sketching.

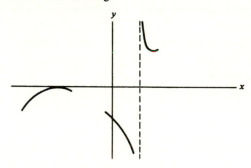

12.10. Completed sketch.

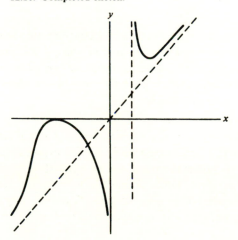

passes through -2. Therefore the function touches the x-axis at this point but does not cross it. Moreover its coefficient is negative, namely $(-2-1)^{-1} = -\frac{1}{3}$. Add this feature to the sketch (fig. 12.9).

(iii) Check the limiting behaviour as $x \to \pm \infty$.

If $|x|$ is very large, say 10^6, then $x + 2 \approx x$ so that in this limit, $y \approx x^2/x = x$. This is called its *asymptotic* form. But $y = x$ is just the straight line through the origin at $45°$ to the x-axis. Therefore, at large distances in the positive or negative direction, the slope of the actual function must tend towards unity. Add this to the sketch.

(iv) Look for any further identifiable features. If none, join the portions and the sketch is complete (fig. 12.10).

Summary

In graph plotting, attend to:

(*i*) Origin, i.e. values of the variables where the axes cross.

(*ii*) Scale. Select the best ratio of length to data unit to give necessary coverage and accuracy.

(*iii*) Choice of functional forms. To find parameters, choose combinations of variables to give a straight line (if possible).

(*iv*) Errors. (See chapter 2.)

In curve sketching, look for principal features, which include

(*i*) singularities,

(*ii*) zeros,

(*iii*) asymptotic behaviour.

EXERCISES

1. What kind of plot would you try if you had data for $y(x)$ over some range and wanted to make an estimate of the limit of y as $x \to \infty$? This is the process of *extrapolation*, in which we try to reach out beyond the range of available data. Interpolation, which fills in the range between data points, is usually much more straightforward.

2. The plot shown in fig. 12.4 was made in order to investigate a further possible source of systematic error, which would depend on h. What might this be, and how might it depend on h? Note the importance of the error bars in the case – no systematic trend can be established, since $h^2 t =$ constant within the error limits. Derive a formula for the total uncertainty of $h^2 t$ in terms of the uncertainties of h and t. If only a single measurement was to be made, what height h should be chosen, to minimise random error?

3. Sketch the following functions:
   (i) $f(x) = x^{-2}(x^2 - 2)^{-1}$
   (ii) $f(x) = x^{-2} \cos x$
   (iii) $f(x) = \frac{1}{2}(e^x - e^{-x})$ ($= \sinh x$, hyperbolic sine)
   (iv) $f(x) = \frac{1}{2}(e^x + e^{-x})$ ($= \cosh x$, hyperbolic cosine)

4. In a gas at constant temperature, the pressure p and volume V are related approximately by the van der Waals equation
   $$(p + A)(V - B) = C,$$
   where A, B, C are given positive constants. This is an improvement upon the ideal gas law given in the text.
   (i) Express pressure as a function of volume and sketch its general behaviour for several values of C. Indicate stationary points, asymptotic behaviour, etc.
   (ii) Show how for given values of A, B, C and arbitrary pressure this corresponds to either (a) a unique volume, or (b) three values of volume.

5. On separate sheets of mm^2 paper draw graphs of $y = 2/x$, and $y^2 = x^2 - 4$ for $-5 \leqslant x \leqslant 5$. Superimpose the graphs with a common origin and verify that with a suitable relative orientation they can be made to agree. Attempt to explain this.

6. The coordinates x and y are related according to the equation $2x^2 - 4xy + y^2 = 1$. Without solving for y (or x) find the general shape of the curve represented.

7. The specific heat of a metal at low temperature is represented by the function of temperature T,
   $$C = aT + bT^3$$
   where a, b are constant positive coefficients. Sketch C against T for positive T. Suggest possible plots to determine a, b from a given set of measurements of C with T.

8. In the theory of radiation there occurs the Planck function,
   $$y = \frac{x^3}{\exp(x) - 1}, \quad x \geqslant 0.$$
   Sketch the behaviour of this function for $x \geqslant 0$, indicating the stationary points. Find approximations for y, (i) for $x \ll 1$, and (ii) for $x \gg 1$.

9. The coordinates of x, y are related by the equation,
   $$x^4 - 6x^2y^2 + y^4 = V,$$
   where V is a fixed parameter. What is the behaviour of y as x becomes indefinitely large? If x is small show that y is approximately given by $y = \pm (V^{\frac{1}{4}} + \frac{3}{2}V^{-\frac{3}{4}}x^2)$. Sketch the expected shape of the x, y curve for various values of V.

10. Two important potential energy functions $V(r)$ are
    (a) the Yukawa potential, describing the interaction between two nucleons
    $$V = (a/r)\exp(-r/r_0)$$
    and
    (b) the Morse potential for the interaction of two atoms
    $$V = V_0[e^{-2(r-r_e)/a} - 2e^{-(r-r_e)/a}].$$
    Sketch the form of each potential function.

13
Differentiation

Wherever there is a function, theory finds a star role for its 'rate of change'. Indeed, relations (differential equations) which involve this quantity are the simplest means the human mind has found to express the principles of such sciences as dynamics, electromagnetism, thermodynamics, and even the statistical laws of chance.

Such an important and practical notion can hardly be abstruse. Although its inventors Newton and Leibniz were innovators of genius, common mortals can easily pick up its use. First of all comes the functional relation: speed versus time; heating-oil consumption versus outside temperature (a steepening curve!); mean river speed versus river width. Sometimes 'continuous' is interpreted loosely, as in the frequency of road repairs versus the mean traffic flow, but the idea remains the same. One variable is selected as 'independent' (time, perhaps), and the other as 'dependent'. The choice is not always compelling – is traffic flow controlled by road repairs or are repairs caused by traffic? The mathematician, passionate for abstraction, denotes the independent variable by plain x, the dependent variable by y and expresses the relation as $y = f(x)$. Actually, the mathematician probably has in mind an algebraic relation of y to x, and might object that the above examples do not constitute 'differentiable' relationships. With a graph or tabulation of $y = f(x)$ there is usually no difficulty in numerically estimating the differentiated function. This is just the local slope plotted (or tabulated) against x. 'Slope' of course is just convenient shorthand for 'rate of change'. There are odd corners of physics where non-differentiable functions are important but here we will cling to the notion that physical variables always vary smoothly.

There is a simple recipe for finding the local slope at any point x. Select a nearby point, $x + \Delta x$ say, where Δx is as small as practicable, and then calculate the difference $f(x + \Delta x) - f(x)$, which will also be small. The ratio of these two small quantities, namely $[f(x + \Delta x) - f(x)]/\Delta x$, is an estimate of the slope; it is a finite number depending slightly upon Δx, and what we really want is its limit as $\Delta x \to 0$.

Such a limit (the *derivative* of $y = f(x)$ with respect to x) could be denoted by $\lim_{\Delta x \to 0} (\Delta y/\Delta x)$. This is cumbersome and so it is shortened to dy/dx, where dy, dx are symbols representing 'infinitesimals' – indefinitely small changes in y, x (the 'ghosts of departed quantities'). This notation really represents a style of thought that has been abandoned in the best mathematical circles, but still remains in the minds of many of us. Alternatively, d/dx may be regarded as an *operator* which acts on y to give its derivative. In another notation the derivative function may be written as $f'(x)$, or in the special case when time is the independent variable, simple $\dot{f}$ (This is the 'fluxion' dot-notation, due to Newton).

The attentive reader will not have failed to notice that these ideas, if not the terminology, have been in use for the last four chapters. Thus from their geometrical definitions (formulae 11.3) $[\cos(x + \Delta x) - \cos x]/\Delta x = -\sin x$, and $[\sin(x + \Delta x) - \sin x]/\Delta x = \cos x$. It follows also from (10.2) that the derivative of $\exp x$ is (unusually but characteristically) itself, $\exp x$.

A practical example illustrates the general procedure of *numerical differentiation*. In fig. 13.1 a smooth curve has been drawn through the nine

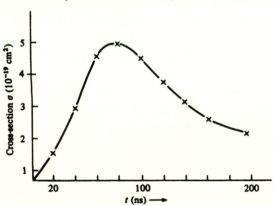

13.1. Experimental data relating to optical absorption in a gas.

experimental points (marked with crosses) of a set of measurements. To some extent the interpolation is subjective, but in any case the experimental errors (not shown) do not allow great precision. The horizontal range is 0 to 200, so that it is practicable to take $\Delta x = 10$, although this will give some error round about the maximum, $x \approx 80$. Actually the units are given as ns (1 nanosecond = 10^{-9} seconds), but this need not come into the calculations. The dependent variable is plotted on the vertical axis (the ordinate) and ranges from 0 to 5. The units are 10^{-19} cm^2 but again do not affect the calculation. Thus the plot can be tabulated as below (table 13.1).

The result is plotted in figure 13.2 with a change of scale. Thus the ordinate unit represents 10^{12} cm^2/s. Note that the derivative is roughly

Table 13.1   *Numerical data corresponding to fig. 13.1.*

Abscissa, x (in units of 10^{-9} s)	Ordinate, $y = f(x)$ in units of 10^{-19} cm^2	$[f(x + \Delta x) - f(x)]/\Delta x$, $\Delta x = 10$ (units 10^{-10} cm^2/s)
30	2.0	0.08
40	2.8	0.08
50	3.6	0.07
60	4.3	0.04
70	4.7	0.01
80	4.8	-0.02
90	4.6	-0.03
100	4.3	-0.03
110	4.0	-0.03

13.2. Derivative $d\sigma/dt$, estimated for the data of Fig. 13.1.

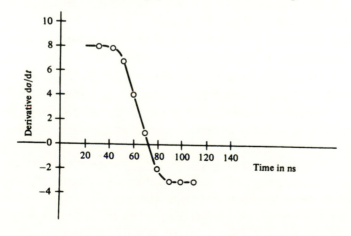

constant as the function starts to climb, then changes sign at the peak and settles down to a (different) constant value.

The same procedure could be repeated for fig. 13.2, to produce the second derivative of the original function, but the uncertainty would by then be considerable. On the other hand, when we use the rules of calculus to differentiate an algebraic functional form, we can take higher derivatives without any such problem. We shall not tabulate the derivatives of the various elementary functions used in this book (see any calculus text) but we give below a reminder of the two principal rules in their use.

(*i*) Products of functions

$$\frac{d}{dx}(fg) = \frac{df}{dx}g + f\frac{dg}{dx},$$
(13.1)

(*ii*) Function of a function

$$\frac{d}{dx}f(y(x)) = \frac{df}{dy}\frac{dy}{dx}$$
(13.2)

A common and useful application of the first derivative is to find the stationary points of a function, which will include its maxima and minima. Since, by definition, the slope is zero at such points, then $f(x)$ is stationary at point $x = a$ if $[f'(x)]_{x=a} = f'(a) = 0$. Thus in the foregoing numerical example, the derivative curve crosses the axis at about 70.2, which gives a rather more accurate value than is obtained by casually inspecting the original experimental points.

For an algebraic example consider the 'oscillatory decay' function $f(x) = \exp(-x)\cos x$. Then $f'(x) = \exp(-x)(-\cos x - \sin x)$, so that since $\exp(-x) \neq 0$, $f'(a) = 0$ gives $\cos a = -\sin a$. The stationary points of the function are therefore given by $\tan a = -1$, the solution of which is $a = (n + \frac{3}{4})\pi$, where n is any integer (cf. fig. 11.1)

If it is not clear whether a stationary point defines a maximum or a minimum, then we must evaluate the *second* derivative. In the above example, $f''(x) = 2\exp(-x)\sin x$ so that $f''(3\pi/4) = \sqrt{2}\exp(-3\pi/4)$ which is positive so that $x = 3\pi/4$ is a minimum. Note that a positive second derivative means an increasing slope and hence an upward curvature. It could happen that $f''(a) = 0$. In this case $x = a$ is called a *point of inflexion* (see chapter 12).

Finally, we note that when derivatives are needed in computations they may be taken by the same straightforward procedure which was applied to experimental data above. Suppose for example that we need to evaluate the derivative of $\tan x$ at $x = \pi/4$. This could, of course be performed by use of the formula $\dfrac{d}{dx} \tan x = \sec^2 x$, but if we wish to do it numerically instead, we may use

$$\left[\frac{d}{dx} \tan x \right]_{x=\pi/4} \approx \frac{\tan\left(\dfrac{\pi}{4} + \Delta x\right) - \tan\left(\dfrac{\pi}{4} - \Delta x\right)}{2\Delta x}. \qquad (13.3)$$

This must be used with a small value of Δx, as fig. 13.3 illustrates, in order to give an acceptable estimate of the limiting value as $\Delta x \to 0$. In chapters 14 and 15 we shall develop a better understanding of this and also see why this particular formula is recommended, rather than the more obvious one

$$\left[\frac{d}{dx} \tan x \right]_{x=\pi/4} \approx \frac{\tan\left(\dfrac{\pi}{4} + \Delta x\right) - \tan\dfrac{\pi}{4}}{\Delta x}. \qquad (13.4)$$

In practice, it is by no means the case that 'the smaller the Δx, the better the answer', at least not indefinitely. The choice of too small a value will incur serious round-off errors, because we calculate the difference between two almost equal numbers. A suitable compromise value must be chosen (and tested) for the purposes at hand, so numerical differentiation, although easy and quick in execution, is an awkward procedure. In the present case, calculating $\sec^2 x$, the derivative of $\tan x$, at the specified point is much the

13.3. Numerical derivative of tan x at x = π/4, using (13.3) as a function of Δx. Note that this tends to the value sec²π/4 = 2 as $\Delta x \to 0$.

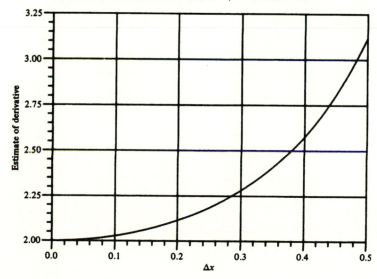

better idea. Similarly, experimentalists try hard to measure the derivative of a physical quantity in some *direct* manner (rather than as in our example above) if it is their prime object of interest. The independent variable x is often 'modulated' in the experiment (varied in a controlled way, e.g. as in a sine wave) and the corresponding variation of the measured property $f(x)$ is picked up and related to the derivative $f'(x)$. This avoids the necessity for numerical differentiation and often makes for substantially improved accuracy.

Summary

First derivative (slope) df/dx or f': zero at stationary points
Second derivative (upward curvature) d^2f/dx^2 or f'': positive/negative at minimum/maximum, zero at a point of inflexion.

Algebraic differentiation: $(fg)' = f'g + fg'$, $(f(y))' = f'y'$.

Numerical differentiation: beware of serious errors (e.g. round-off) due to subtraction in formula.

EXERCISES

1. If round-off error is roughly one part in 10^6, how small an interval may be used in the numerical derivative (13.3) without incurring uncertainty, due to round-off, greater than 1%? If this value is used, does this mean that the derivative is correct to within 1%? How could you investigate this, without using the known properties of the function $\tan x$?

2. Draw sketches illustrating the form of the following functions and their first and second derivatives: (a) $\sin x$ (b) $\ln x$ (c) $x^{\frac{1}{2}}$ (d) $\exp(x^{-1})$ (e) $[x(1-x)]^{\frac{1}{2}}, 0 \leqslant x \leqslant 1$.

3. 'The government has announced a slowing of the rate of increase of inflation.' Defining the value of the pound as a function $f(t)$, identify the derivative which is the subject of this BBC News announcement.

4. A billiard ball rolls across the table in a straight line, rebounds from the cushion and returns, coming to rest. Draw a careful sketch of position x, velocity $\dot{x}$ and acceleration $\ddot{x}$ as functions of time.

5. Two different nuclear species A and B have approximately the same half-life λ. Show that if they are initially present in equal numbers in a given sample, the difference between the number of A and B nuclei at a later time t is proportional to $te^{-t/\lambda}$. When does this function take its maximum value?

6. If x and y are related according to $2x^2 - 4xy + y^2 = 1$, show that
$$\frac{dy}{dx} = \frac{x - y}{x - \frac{1}{2}y}.$$
Use this result to further discuss exercise 6 of chapter 12.

7. Sketch the following functions and find their stationary points, identifying them as maxima, minima or points of inflexion:
   (i) $10x^2 \exp(-x^2)$, $x \geqslant 0$;
   (ii) $x \exp(-x^2)$, all positive and negative x;
   (iii) $10 \exp(-2x) - 2 \exp(-x)$, $x \geqslant 0$.

8. Sketch the function $x(1 + x^4)^{-1}$ for all positive and negative x. Given that a function $y = f(x)$ satisfies the equation

$$\frac{dy}{dx} = \frac{x}{1 + x^4}$$

discuss the behaviour of $f(x)$ and sketch its possible shape.

9. For each of the two interaction potentials $V(r)$ defined in exercise 10 chapter 12, derive a formula for the position of the minimum of the potential.

10. The attractive force between two atoms is given by the derivative of the interaction potential $V(r)$. Given such interaction potentials and a particular configuration of a molecule (distorted from equilibrium), describe how the total force on each atom could be calculated.

14
Approximations

'Fine tuning' is an expression much in the mouths of economists and planners, meaning (some would say) fiddling adjustments. It is genuinely descriptive of measuring instruments, such as the micrometer, which have a coarse setting supplemented by a vernier scale to provide final accuracy. In the original context, the frequency of a radio station or signal on some wave band would be first located roughly, then 'fine tuned' with a more sensitive control to find the maximum signal. For this to succeed all that is required is that a dependent quantity (e.g. loudness of a signal) depends smoothly on a continuous variable (radio frequency). Moreover this dependence, what-

14.1. Snell's Law, $\sin i = \mu \sin r$

ever its form on a broad scale, tends to become linear for short-range variations.

In many physical situations there is a natural operating point. For example, in the refraction of light at a plane optical interface, air/glass say, the directions of the incident and refracted rays are related by Snell's law. This states that $\sin i / \sin r = \mu$ (Greek *mu*), which is a constant (see fig. 14.1). Taking the angle i (in radians) as the variable x, the angle r is given by the function $f(x) = \sin^{-1}(\mu \sin x)$. But commonly the incident ray is nearly normal to the surface, i.e. $x \ll 1$, and in chapter 11 it was shown that for $x \ll 1$, $\sin x \approx x$. Since μ is not much bigger than 1 it follows that for such rays $f(x) \approx \mu x$.

In the above example the function $f(x)$ was approximated near the value $x = 0$ where it happened that $f(0) = 0$. But consider another example. A photographic process is very sensitive to temperature. Suppose that the development time t (for a standard contrast) varies with the developer temperature T according to the function

$$t = f(T) = 0.1 \exp\left(\frac{0.9 \times 10^3}{T}\right) \tag{14.1}$$

with an appropriate choice of units for time and temperature. If the dependent variable t is plotted against T (using the results of Chapter 10), it takes the form of the upper curve shown in fig. 14.2.

Suppose now the standard operating condition is at room temperature, T_0 say, with a corresponding 'standard' time $t_0 = f(T_0)$. Then in the vicinity of T_0, the function is well approximated by the tangent PQ. On this line, a small temperature change ΔT is proportional to a small time change, Δt. Their ratio is simply the slope of the line PQ. Thus over a limited range, (14.1) is approximated by

$$\Delta t = f' \Delta T, \tag{14.2}$$

14.2. Variation of exposure time with temperature in a photographic process.

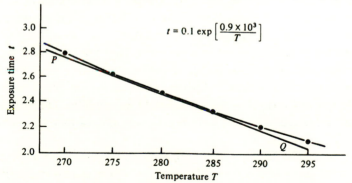

where the coefficient f' is the (negative) slope at the operating point.

Some particular linear approximations of this kind which recur frequently in physical science are listed below:

$$
\left.
\begin{aligned}
&\sin x \approx x \\
&\tan x \approx x \\
&\exp x \approx 1 + x \\
&\log(1 + x) \approx x \\
&(1 + x)^{-1} \approx 1 - x
\end{aligned}
\right\} \qquad \text{for small } x \qquad (14.3)
$$

Such a linear approximation can become inadequate. This may be because the function is curving too rapidly at the point of interest so that the tangent approximation leads to unacceptable errors. The next chapter provides the remedy, for (14.3) consists merely of the first two terms of the Taylor series which is discussed there. We can take as many higher powers of x as we need. Many terms may be necessary if we try to approximate a function over a wide range. Thus if the error associated with our linear approximation is unacceptably large, we have two options – to incorporate more terms or to shrink the range in which the approximation is used. Practical computer methods often use the latter approach for reasons of simplicity and also because we may encounter round-off errors due to the cancellation of terms.

Suppose that we shrink the range ($x - h$ to $x + h$) over which such an approximation is used, in order to 'tighten' it. How fast will the error decrease? If the first neglected term involves x^n, we may expect the error to be dominated by this, at least eventually, as $h \to 0$. We say that the error is of order h^n, or $O(h^n)$. This means 'varies as' h^n (at worst). Such statements are very helpful in trying to control errors, even if no precise bounds on the errors are given. If the error is $O(h^4)$ for example, and we wish to reduce the error in a given calculation by a factor of 10^2, then the range should be reduced by a factor of $10^{\frac{1}{2}}$. Much the same reasoning is involved in dealing with errors in numerical integration (chapter 17) and in the numerical integration of differential equations (chapter 19).

Functions can be approximated in other ways. They are often written as linear combinations of some simple standard functions. In a sense this is what a power series is, but instead of using powers of $(x - a)$ we could use, for example, sine and cosine functions of different wavelengths. The Fourier series (chapter 24) expresses this idea, but is only one of many possible choices.

Note the distinction between such approximations, which can be made more and more accurate in a controlled way, and mere guesswork, which should not really be called approximation.

Summary

A function may be represented over a limited range by a linear approximation. If this is inadequate, higher powers may be included. The approximations $\sin x \approx x$, $\tan x \approx x$ (for small x) are particularly useful.

EXERCISES

1. Plot the function $f(x) = 1 + x + \frac{1}{2}x^2$ for $-1 \leqslant x \leqslant 1$ at intervals of 0.2. Compare the graph with that of $\exp x$, and evaluate the relative errors at $x = +1$.

2. Use the approximations (14.3) (without the aid of a calculator) to obtain quick answers to the following: $(80)^{\frac{1}{4}}$, $(1000)^{\frac{1}{10}}$, $\ln 21$.

3. Use the logarithmic approximation and the result $2^{10} = 1024$ to find an improved value for $\log_{10} 2$.

4. How would you approximate $(x-1)^{-1} \ln x$ around $x = 1$? This is an example of an 'indeterminate' form which cannot be evaluated at the point in question by simply evaluating its individual factors in the usual way. Figure 14.3 is a plot of this function, which will help you to see the form of the required result.

5. The bob of a pendulum of length l is displaced horizontally by a distance x. Find the corresponding vertical displacement, to second order in x, by using trigonometry and an approximation for $\sin \theta$. (This can also be done by expanding $\cos \theta$ to second order, as in the next chapter.)

14.3. Full line: $\ln x$ Dashed line: $(x-1)^{-1}\ln x$

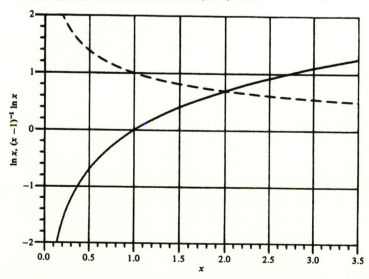

6. The extension of a spring is found to be well approximated by $x = 20F - 5F^2 - F^3$ for a force F of magnitude $0 < F < 1$, in certain units. What force F_0 corresponds to $x = 10$? Obtain the solution of the relevant equation by *linearising* it. To do this, make a reasonable guess $F_0^{(1)}$ for F_0. Substitute $F_0 = F_0^{(1)} + y$ in the equation; keep only terms up to those linear in y (i.e. neglect y^2, y^3); solve for y. This is a powerful method of approximation in many branches of applied mathematics. Note that successive approximations can be used, by using your result as the starting value for another linearisation, and so on. Optional: How would you expect the error to vary as successive approximations are taken? Test your result for this example using, say, three successive approximations.

7. Stirling's approximation for factorials is
$$n! \approx (2\pi)^{\frac{1}{2}} n^{n+\frac{1}{2}} \exp(-n) \tag{14.4}$$
but physicists generally use it in the form
$$\log n! \approx n(\log n - 1) \tag{14.5}$$
Show that (14.5) is consistent with the more accurate formula, apart from terms which increase more slowly than n, as $n \to \infty$. Check the accuracy of the approximation (14.5) for $n = 5$ and $n = 2$ by numerical calculation. A derivation of (14.5) is suggested in exercise 8 of chapter 17.

8. Examine the behaviour of $\tan x$ around $x = \pi/4$ and hence explain why the formula (13.3) was preferred to (13.4) in chapter 13.

9. When a metal filament is heated, electrons are emitted from it at a rate given, as a function of a temperature T, by
$$I = AT^{\frac{1}{2}} \exp(-B/T) \quad \text{(Richardson's equation)}.$$
where A and B are positive constants. Sketch this function of T. Derive an equation for the temperature T_0 at which the rate of change of I with temperature is a maximum.

10. Suppose that for ease of computation we require a quadratic approximation $f(x) = A + Bx + Cx^2$ to the function $\sin x$, to be used in the range $0 \leqslant x \leqslant \pi$.
   (a) Find A, B and C by solving the set of equations obtained from $f(x) = \sin x$ at $x = 0, \pi/2, \pi$. Find the maximum error in the stated range.
   (b) Since the above is clearly arbitrary, can you suggest any criterion for a 'best' choice of A, B, and C?

15

Power series and Taylor's expansion

The heart of most qualitative description of physical systems is the mathematical function, expressing the dependence of one variable quantity upon another. The named (or 'special') functions dealt with so far – the exponential, logarithm, cosine, and related functions – constitute a basic set of tools for describing functions. Their use can be extended enormously by combining them in various ways, so that in principle there is very little functional behaviour that they cannot represent. But for many purposes it is more convenient to specify additional functions. Often these are solutions of differential equations (chapters 18–20). By such means, further important named functions are introduced into the mathematical vocabulary, such as Bessel, Legendre, hypergeometric, and so on. The definition and systematisation of a whole class of named and related functions is often pursued through their representation by *Taylor (or MacLaurin) series*.

But such *power series* have more direct practical uses for the physicist, namely to approximate functions in numerical calculations when the linear approximation of the last chapter becomes inadequate but only a few additional powers are needed. Sometimes experimental data is confined to such a range and the goal of measurement is to find the first few coefficients of a power series.

Suppose then that we require values of a function $f(x)$ close to a point $x = a$. The appropriate power series expansion is

$$f(x) = c_0 + c_1(x - a) + c_2(x - a)^2 + \cdots, \tag{15.1}$$

as was mentioned in chapter 14.

The coefficients c_m are constants, and clearly $c_0 = f(a)$, but how are the values of the higher coefficients related to $f(x)$ itself? The answer is provided by *Taylor's theorem*,

$$c_n = \frac{f^{(n)}(a)}{n!} \tag{15.2}$$

so that

$$f(x) = \sum_{n=0}^{\infty} \frac{f^{(n)}(a)}{n!}(x - a)^n \tag{15.3}$$

where $f^{(n)}(a)$ denotes the nth derivative of f at $x = a$, i.e. (n) represents n primes. Check that this makes sense, at least for the first few terms, by repeatedly differentiating the right-hand side of (15.1) and taking the limit $x \rightarrow a$.

Often only the first two terms are required, giving the linear approximation already discussed in chapter 14.

It is clearly necessary that $f(a)$ and the derivatives $f^{(n)}(a)$ exist and are finite. As a counter-example, $f(x) = x^{\frac{1}{2}}$ cannot be expanded in powers of x because all the derivatives diverge as $x \rightarrow 0$. Existence of the derivatives at $x = a$ is not, strictly speaking, enough to guarantee the validity of Taylor's expansion over a finite range of x – but exceptions are rare in physical science.

Assuming the validity of the theorem and that the coefficients of $f^{(n)}(a)$ are known, or capable of evaluation, what is the meaning of 15.3? The representation of the function $f(x)$ over a range of values of x by such an infinite series of powers requires that partial sums (i.e. terminated series) eventually converge to a limit (the 'sum' of the series). A sufficient number of terms will then provide an approximation to $f(x)$ to any required accuracy. The actual number of terms required may depend upon the value of x and indeed in special cases may become indefinitely large as x approaches some critical value. The series representation must not be 'pushed too far'.

This question of convergence can be illustrated with two examples, both being series commonly occurring in physical contexts. The first of these is the geometric series, easily visualised in terms of a cake-cutting problem. A round cake is to be equally shared among three people – direct trisection is difficult, so do it in steps. First divide it into four equal parts (easy) and share out three of them. Divide the remaining quadrant into four and share out three sections to give an extra $(\frac{1}{4})^2$ per person. It is clear that dividing the left-over section each time by four and sharing three of the pieces, eventually (crumbs permitting) the cake is fully and equally shared, i.e. trisected. There is nothing mathematically special about trisection so that with an obvious generalisation from ($\frac{1}{4}$) shares to x-shares, with $x < 1$, division by $(1/x) - 1$ is achieved with an infinite number of steps. Thus

$$\lim_{N \to \infty} \sum_{n=1}^{N} x^n = \left(\frac{1}{x} - 1\right)^{-1}. \tag{15.4}$$

The left-hand side is a geometric series commencing $x + x^2 + x^3 + \cdots$; with a slight re-arrangement, the formula can be more conveniently written

$$\sum_{n=0}^{\infty} x^n = (1 - x)^{-1}. \tag{15.5}$$

It should shortly become clear that (15.5) in fact holds for $-1 < x < 1$, and not outside this range.

Equation (15.5) is easily checked for consistency with Taylor's theorem since the first derivative of $(1 - x)^{-1}$ is $(1 - x)^{-2}$, the second derivative is $2(1 - x)^{-3}$, the third $6(1 - x)^{-4}$, and so on. Putting $x = 0$ for the expansion about $x = 0$ gives $f^{(n)}(0) = n!$ which, inserted into (15.1), cancels the already present factorials to give (15.5).

As x approaches the value 1, the function $(1 - x)^{-1}$ increases indefinitely, and from the series point of view, more terms must be taken to complete the sum. To see this directly, suppose the series is terminated at the Nth term. Then the *partial* sum is given by

$$1 + x + x^2 + \cdots + x^N = (1 - x)^{-1} - x^N (1 - x)^{-1} \tag{15.6}$$

as may be shown by straightforward algebra. The second term on the right-hand side is obviously the *remainder* – the discrepancy between the partial and infinite sums (cf. (15.5)). Only for $|x| < 1$ does the remainder go to zero as $N \to \infty$, as required for (15.5) to be correct (or even meaningful). The point $x = 1$ is not covered by (15.6) and needs special (but trivial) consideration. Note also that in proving convergence of a series, it is *not* sufficient to observe that its *terms* decrease – it is the remainder that matters.

The point $x = 1$ where the function diverges (goes to infinity) is said to lie on the 'circle of convergence'. This terminology really belongs to complex variable theory (chapter 22). Diametrically opposite to $x = 1$ is $x = -1$ which is also the point beyond which the series no longer converges.

The size of the circle of convergence is determined by the nearest 'singularity', by $x = 1$ in the present case. Possible forms of singularity are divergences ($f \to \pm \infty$) or discontinuities of f or its derivative. The possibilities are wider than this, but we shall not attempt a precise definition.

Turning now to the exponential function $f(x) = \exp x$, this may be defined as the solution to the differential equation $df/dx = f$ with the initial condition $f(0) = 1$ (see chapter 10). It follows that $d^2f/dx^2 = f$, $d^3f/dx^3 = f$, and so on, so that at any point $x = a$, $f''(a) = f(a)$. With this common value for the coefficients, (15.3) becomes

$$f(x) = f(a)[1 + (x - a) + 1/2!(x - a)^2 + \cdots]. \tag{15.7}$$

Taking $a = 0$, and using the initial condition, the exponential series becomes

$$f(x) = \exp x = 1 + x + x^2/2! + x^3/3! \cdots \tag{15.8}$$

The convergence of 15.8 (for all x) is confirmed by taking the ratio of the $(N + 1)$th term to the Nth, namely x/N. Comparing this with the constant ratio x, for the geometric series, it is concluded that (15.8) must always converge. No matter how large x may be, eventually x/N becomes less than 1, and the terms of the series decrease faster than those of a convergent geometric series.

Similarly, the expansions of $\sin x$ and $\cos x$ are

$$\sin x = x - \frac{1}{3!}x^3 + \frac{1}{5!}x^5 \ldots \tag{15.9}$$

$$\cos x = 1 - \frac{1}{2!}x^2 + \frac{1}{4!}x^5 \ldots \tag{15.10}$$

Figures 15.1–15.4 illustrate the way in which the partial sums of such series converge to the values of the corresponding functions. Note in

15.1. Magnitude of individual terms in the power series which represents exp 10, i.e. $10^n/n!$

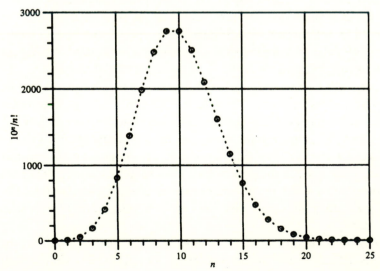

15.2. Partial sum of the series for exp 10, including terms up to $10^n/n!$ as a function of n.

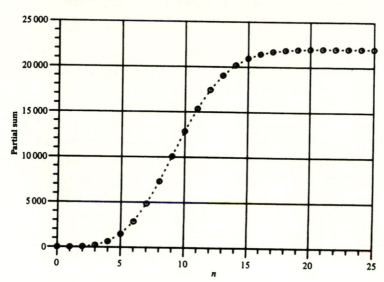

15.3. The function exp x (full line) is compared with the functions defined by partial sums up to x^{10} (- - - -) and x^{15} (············).

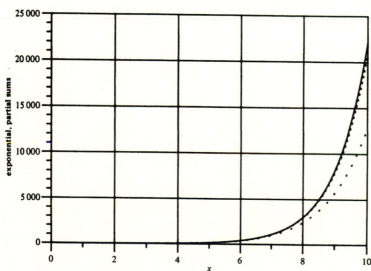

particular that the case of sin x is much more demanding. Successive terms in the series partially cancel, due to their alternating signs, so that many more terms are required than in the case of exp (x), for the same *relative* accuracy.

In considering questions of convergence of series it is also worth remembering that any series whose terms alternate in sign, and decrease successively in magnitude must converge (Abel's theorem).

A nice example of a power series in physics is provided by the specific heat of a solid. In elementary courses, this is often treated as a constant, but it is not – indeed its variation with temperature played a key role in showing the necessity for quantum theory, in the early years of this century. This variation is most striking and significant at low temperatures, since the specific heat C goes to zero at the absolute zero of temperature T. In insulators, it is found experimentally that

$$C = cT^3 + \text{(higher powers of } T) \tag{15.11}$$

The relevant theory provides only rather messy solutions for C (arising from the vibrations of atoms) but using Taylor's theorem, the first few terms can be identified and compared with (15.11). In metals, on the other hand, a *linear* dependence on T is observed (provided the sample is not super-conducting) so the vibrational theory is inadequate. It turns out that the contribution of the free electrons in a metal, similarly expanded, gives the required linear term.

15.4. The function sin x (full line) is compared with the functions defined partial sums up to x^5, x^{10}, x^{15}, x^{20}

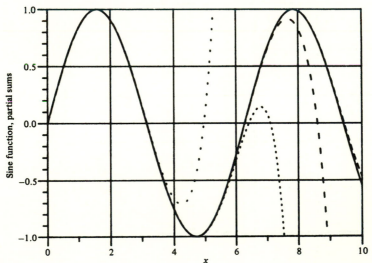

Many of the linear relationships previously mentioned in earlier chapters (even matrix relationships!) are really just the first terms of power series. Suitable experiments may expose their limitations, and demand power series. Hence, for example, the modern subject of 'non-linear optics', brought about by the high intensity of laser light.

In physical problems, functions do sometimes arise which do not qualify for a Taylor expansion. For example, the radial (inward) force describing the gravitational attraction due to the Earth can be functionally expressed to vary with distance x from its centre as

$$f(x) = \begin{cases} Mg(x/R), & 0 \leqslant x \leqslant R \\ Mg(x/R)^{-2}, & R \geqslant x \end{cases} \quad . \tag{15.12}$$

It is apparent that $f(x)$ cannot be expanded in a Taylor series about $x = R$ because df/dx is discontinuous there, as fig. 15.5 shows. Also, note the danger if we attempt to expand about a point beyond $x = R$ using derivative coefficients obtained from (15.3). For $x < R$, the resulting series, which describes not $f(x)$ as defined above but rather the continuation of $Mg(x/R)^{-2}$, is indicated by the dotted line in fig. 15.5. So the point at which a series expansion must be abandoned is not always the point at which it actually diverges.

A somewhat deeper problem arises in magnetism. It is known, for example, that a weak magnetic field B produces in many materials a magnetisation proportional to B (dia-or paramagnetism). In stronger fields the magnetisation may show a saturation effect, described mathematically by correcting the linear term with a (negative) cubic term i.e. $\propto B^3$. Still

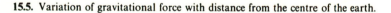

15.5. Variation of gravitational force with distance from the centre of the earth.

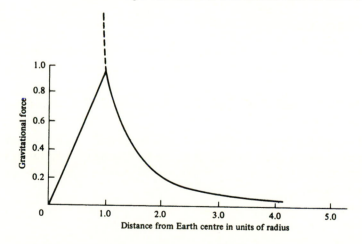

higher fields may require a B^5 term and so on, suggesting that an accurate theory would lead to a Taylor series, just as in the quantum theory of specific heats mentioned above. This turns out to be true, but not the whole truth. As well as the expected power series, detailed calculations (for certain metals) show that additionally the magnetisation contains a term dominated by the factor $\exp(-a/B)$ with $a > 0$. Such a highly singular term cannot be represented by a Taylor series starting from $B = 0$, and is a warning that powers can be abused! A similar singularity occurs in the BCS (Bardeen, Cooper, Schrieffer, 1957) theory of superconductivity for which the authors were awarded the Nobel Prize. Its lurking presence is a probable reason for the failure of many earlier attempts on that problem.

However we should not be too alarmist about such possibilities. In the great majority of cases we deal with 'well-behaved' functions, for which the Taylor series works well.

Summary

The functions which occur in physical science can be usually expanded about a chosen point as a Taylor series

$$f(x) = \sum_{n=0}^{\infty} \frac{f^{(n)}(a)}{n!} (x-a)^n$$

In general this will only converge in a certain range defined by a radius of convergence $x = a \pm R$ where R is the radius of convergence.

At occasional *singular* points, the expansion is not possible; moreover, these set limits on the range over which it converges, if made with respect to a nearby point.

EXERCISES

1. Evaluate $(1-x)^{-1}$ to three decimal places for $x = 0.5$ by series expansion about $x = 0$. Try the same procedure for $x = 2$ and comment on what you find.

2. Give the first three terms of the series expansions of $\sin x$, $\cos x$, $\tan x$, $(1+x)^{\frac{1}{2}}$.

3. Show that the derivative of $\exp(-1/x)$ is zero for all n at $x = 0$. (The function itself is given the value zero there.) This shows that Taylor's theorem must exclude this case, since it would otherwise imply $f = 0$ over some interval.

4. Using (15.8), check that the series expansion is consistent with $\exp(x)\exp(y) = \exp(x+y)$

by substituting the series expansion for each exponential (up to the third power).

5. Sum the series $1 - \frac{1}{2} + \frac{1}{3} - \frac{1}{4} \ldots$ on a hand calculator, up to the 20th term. Can you estimate the accuracy of this summation as an approximation to the infinite sum without knowing the exact result for the latter? Comment on the case in which the signs are all positive in such a series.

6. The function $y = (x + 2)^2/(x - 1)$ was said in chapter 12 to have the asymptotic form $y \approx x$, as $x \to \infty$. Find a more accurate description of the asymptotic form, containing three terms, by expanding in powers of x^{-1}.

7. The exact sum of the series in exercise 5 is $\pi/4$. Compare the merits of this, as a means of estimating π, with those of the following alternative:
$$\pi/6 = \frac{1}{2} + \frac{1}{2.3.2^3} + \frac{1.3}{2.4.5.2^5} \cdots$$

8. Derive the series representation of $\ln(1 + x)$ in powers of x. Within what range of x will it converge? (Illustrate your result by some numerical calculations.)

9. The interaction between two atoms is described by the Morse potential as in chapter 10, exercise 10. What quadratic approximation represents the potential in the vicinity of the equilibrium separation?

10. In a ballistic galvanometer, a beam of light is reflected from a mirror on to a flat screen, the rotation of the mirror and hence the angular deflection of the beam being proportional to the current I which is to be measured. The beam is projected on to a flat screen and a linear scale is used, which is adjusted to give correct readings for $I = 0\,A$ (zero deflection) and $I = 1\,A$ ($10°$ deflection). By roughly how much is it in error for $I = 4\,A$? (Do not use trigonometric tables.)

16
Partial differentiation

In the expression of physical laws, the relationship of two quantities often involves a parameter (enigmatically defined in some dictionaries as a constant which can vary). Thus in the photographic process the image density depends upon the light intensity for each set exposure time. A complete record of the dependence of the density upon the light intensity I and exposure time t comprises a set of graphs of D against I for a range of values of the parameter t. But equally the intensity I could be taken as a 'parameter' labelling a set of graphs of D against t.

16.1. Computer plot of a function $D(I, t)$

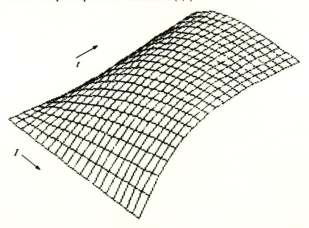

A pictorial alternative to a set of graphs is a surface which relates one variable (height) to two others (plane coordinates). It can be physically modelled by taking silhouettes of the D/I (graphs in cardboard, say) for a set of values of t and pasting them together in sequence. A surface representing D as a function of the two variables I and t might then look like the perspective computer plot in fig. 16.1. For practical use, contour plots provide equivalent information. Taking I and t as Cartesian coordinates, then along a contour curve the density D remains constant, the actual value providing a label. Figure 16.2 provides contours for the surface of fig. 16.1 with D ranging as shown.

The variation of D with I for some fixed t is obtained from fig. 16.2 by

16.2. Contours corresponding to the function shown in fig. 16.1.

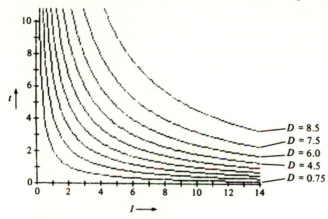

16.3. Plot of D against I for $t = 5$.

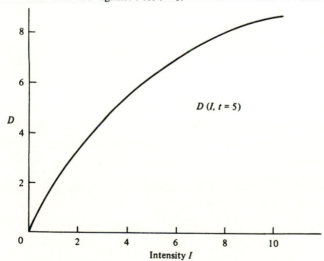

drawing a line parallel to the I-axis at the required t-value and reading off the I-values of the intersections with the successive D-contours. Thus for $t = 5$, points on the $(D/I)_t$ graph are obtained as shown in fig. 16.3. A similar procedure, but fixing I and reading off t-values, provides points on the $(D/t)_I$ curve.

The functional dependence of D upon both I and t is mathematically expressed as $D = F(I, t)$, where $F(I, t)$ provides an algebraic or numerical (graphical) recipe for finding D from a pair of values. Fixing t and using the $D(I)_t$ graph (or the contours) then, as in chapter 13, the rate of change (or slope) may be obtained at each I point. It is denoted by

$$\left(\frac{\partial D}{\partial I}\right)_t$$

and read as the partial derivative of D with respect to I, *keeping t constant*. Holding I constant, the partial derivative of D with respect to t appears as $(\partial D/\partial t)_I$. The brackets and subscript may be omitted in many cases, when it is clear what other variable is being kept constant. The symbol ∂ denotes 'partial' and must not be confused with d. Generally speaking, $(\partial D/\partial I)_t$, $(\partial D/\partial t)_I$ depend on both values t, I, and they give the *slope* of the curves defined by D in the two directions of constant t and I.

Higher derivatives are denoted by $\partial^2 D/\partial I^2$, $\partial^2 D/\partial I \, \partial t$ etc.

The mathematical rules of partial and 'ordinary' differentiation are similar – it is only necessary to keep firmly in mind the variable (or variables) being held constant.

As a second example of the same sort, suppose z is the height at each point (x, y) on a map of some terrain. If we plot a course across this terrain, it can be described by making y a function of x, or making both a function of t, say, which might be time. If we ask the question, how fast does z vary with t (i.e. what is the rate of climb by the mountaineer), the answer is

$$\frac{dz}{dt} = \frac{\partial z}{\partial x}\frac{dx}{dt} + \frac{\partial z}{\partial y}\frac{dy}{dt} \tag{16.1}$$

and is rather like the chain rule (13.2), except that there are two terms because the dependence of z upon t is expressed through two intermediate variables. The following type of diagram is useful in illustrating such a rule. The arrows denote functional relationships: each term in (16.1) represents a chain of such relationships.

$$\tag{16.2}$$

Note that (16.1) contains both partial derivatives and ordinary derivatives. Why?

There are many possibilities for more complicated chain rules, similar to (16.1), particularly when more variables are introduced. With the help of a diagram like (16.2), you should be able to write down (or at least understand) any required rule.

For example, we may consider y to be a function of x (instead of introducing t) and ask for dz/dx. This refers to the variation of z due to its *direct* dependence on x and its *indirect* dependence, through y. The appropriate diagram is

$$z(x, y) \overset{\displaystyle y(x)}{\underset{\displaystyle x}{\longleftarrow}} \qquad (16.3)$$

Again we follow the arrows from x to z and write

$$\frac{dz}{dx} = \frac{\partial z}{\partial x} + \frac{\partial z}{\partial y}\frac{dy}{dx}. \qquad (16.4)$$

In this context the left-hand side is called a *total* derivative, distinguishing it from $\partial z/\partial x$.

Lastly, we might ask for $(dy/dx)_z$, which gives our mountaineer his bearing when he traverses along a contour, at constant height. This is a little more tricky, and we shall need to reintroduce the subscripts to clarify it. The equation for a contour is just

$$z(x, y) = \text{constant.} \qquad (16.5)$$

We can also think of this as defining a curve $y = y(x)$ and, remembering its definition, the derivative of this function can be written as $(dy/dx)_z$. Now we can use the previous rule (16.4), applied to this special case, in the form

$$\frac{dz}{dx} = \left(\frac{\partial z}{\partial x}\right)_y + \left(\frac{\partial z}{\partial y}\right)_x \left(\frac{\partial y}{\partial x}\right)_z. \qquad (16.6)$$

This is *zero* because we are referring to the derivative of z with respect to x along a contour. Rearrangement of (16.6) then gives the required relation

$$\left(\frac{\partial y}{\partial x}\right)_z = -\left(\frac{\partial z}{\partial x}\right)_y \bigg/ \left(\frac{\partial z}{\partial y}\right)_x. \qquad (16.7)$$

One can derive these formulae by approximating the local form of the function $z(x, y)$ by a plane, as in fig. 16.4, which corresponds to the relation

$$\Delta z = \left(\frac{\partial z}{\partial x}\right)_y \Delta x + \left(\frac{\partial z}{\partial y}\right)_x \Delta y \qquad (16.8)$$

between increments of Δz, Δx and Δy. If these increments are associated with the time increment Δt, division of (16.8) by Δt leads to (16.1).

Up to this point we have assumed the existence of some function such as $z(x, y)$ and have considered the effect of increments of its variables. In thermodynamics we encounter relations between increments which are like (16.8) and the question arises – can we attribute them to some function, playing the role of z? That is, can we *integrate* such a relation?

Let us look at a specific example. When a fixed amount of air in a container is allowed to absorb measured amounts of heat under various conditions of pressure p and volume V it is found that the heat absorbed is given by

$$H = c_1 p \Delta V + c_2 V \Delta p. \tag{16.9}$$

Here c_1 and c_2 are constants – for air, $c_1/c_2 \approx 1.4$. The temptation is to write the left-hand side as ΔQ, the 'increment of the heat contained in the sample', but this should be resisted although the notation is sometimes used, somewhat confusingly, *without* such a meaning. To see why this is so, let us calculate the heat absorbed when we proceed from one $(p-V)$ state to another by two different routes.

Suppose the initial thermal state is $p = 1$, $V = 1$ (in unspecified units), while the final state is $p = V = 2$. Proceed at first with p fixed at $p = 1$, expand V from 1 to 2 requiring an absorbed heat of c_1 units. Complete the required change at the fixed final volume $V = 2$ by means of a pressure rise from $p = 1$ to 2, requiring $2c_2$ units, making in all a heat absorption of $c_1 + 2c_2$. But suppose the order of operations is reversed. Then the initial

16.4. The inclined plane is a local approximation to a function $z(x, y)$.

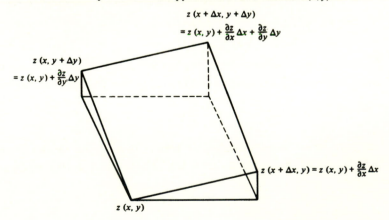

pressure rise (at $V = 1$) requires a heat c_2 while the final expansion requires $2c_1$ units, making a different total, namely $2c_1 + c_2$. Finally, suppose V and p are allowed to rise simultaneously at the same incremental rate, say $V = p$ $= 0.1$ i.e. ten increments each. Then (16.4) gives the sum

$$H \approx (c_1 + c_2)[1.0 + 1.1 \cdots + 2.0] \times 0.1$$

so that, approximately, $H = 1.65(c_1 + c_2)$. This route then gives yet another value for the absorbed heat. All three routes, denoted by a, b, c (in the above order) are shown the p–V diagram of fig. 16.5.

It is evident from these results that the incremental relation (16.9) leads to a net change in absorbed heat which is dependent upon the incremental sequence, or more simply, upon the route or path followed in the p–V diagram. The formula (16.9) is fundamentally different from (16.8) in which a net change $z_2 - z_1$ depends only upon the initial and final points. It follows that the absorbed heat cannot be regarded as a function of p and V. This was a conclusion that physicists found very hard to swallow for many years. It means that there is no such thing as the 'heat contained in a system'. Instead, we must talk about its energy, heat being energy transferred without work. With such a definition, (16.9) can be derived from the law of conservation of energy. Mathematically, the warning is that incremental (or differential) relations such as (16.9) cannot always be integrated! The same point will recur in the theory of potentials – chapter 31. We shall then see that there are important cases in which incremental relations *can* be integrated, but they are best discussed in the language of scalar and vector fields.

16.5. Alternative routes between two points in the p-V diagram.

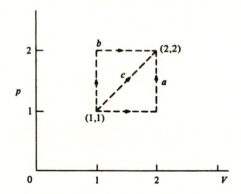

Summary

A function of two variables, $z = f(x, y)$, can be represented as a surface with z as the height corresponding to the coordinate point (x, y). A contour is a curve of constant height.

The partial derivative $(\partial z/\partial x)_y$ is found by differentiating z with respect to x, keeping y constant.

The incremental change Δz due to increments $\Delta x, \Delta y$ is given by

$$\Delta z = \left(\frac{\partial z}{\partial x}\right)_y \Delta x + \left(\frac{\partial z}{\partial y}\right)_x \Delta y.$$

Also,

$$\left(\frac{\partial y}{\partial x}\right)_z = -\left(\frac{\partial z}{\partial x}\right)_y \Big/ \left(\frac{\partial z}{\partial y}\right)_x.$$

In thermodynamics incremental relationships occur which cannot be integrated to find such a function.

EXERCISES

1. Use the contours in fig. 16.2 to calculate D as a function of t for $I = 4.5$. By interpolation to a smooth curve, evaluate $(\partial D/\partial t)_{I=4.5}$ between $t = 0$ and 13.0.

2. Assuming the dependence of D upon I and t to be of an exponential form use the contour data of fig. 16.2 to obtain a formula for the function $D(I, t)$.

3. Sketch the form of the surface $z = 1 - x^2 - y^2$ in the range $0 < x, y < 1$. Plot the contours for $z = 1.0, 0.8, 0.6, 0.4, 0.2$. Obtain the partial derivative $(\partial y/\partial x)$ by (i) solving for y in terms of x, z, (ii) using the formula (16.7). Show that these expressions agree.

4. Sketch the surface $z = x^2 - y^2$, and plot the contours for $z = 0.1, 0.3, 0.5$. If new independent variables are chosen, namely $u = x + y$, $v = x - y$, calculate the partial derivatives $(\partial z/\partial u)_v$, $(\partial z/\partial v)_u$.

5. Using the van der Waals gas equation (chapter 12, exercise 4) with C replaced by RT (denoting the gas constant times the temperature), evaluate $V^{-1}(\partial V/\partial p)_T$ in terms of V, T. Sketch the p, V contours for various T values and discuss their behaviour near $V = B$.

6. Find (by guessing if necessary) functions F which satisfy the following partial differential equations

   (i) $\left(\dfrac{\partial F}{\partial x}\right)_y = 2x + y$, $\left(\dfrac{\partial F}{\partial y}\right)_x = -3y^2 + x$

   (ii) $\dfrac{\partial^2 F}{\partial x^2} = 0$, $\dfrac{\partial F}{\partial y} = F$

   (iii) $\dfrac{\partial^2 F}{\partial x \partial y} = 0$

16.6. Apparatus commonly used to demonstrate equilibrium of three forces.

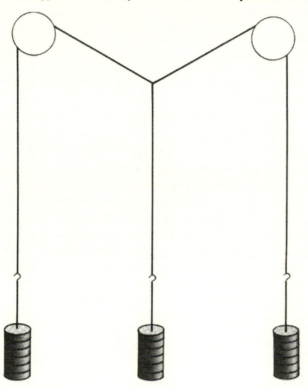

16.7. Function of two variables describing reflection from an optical component.

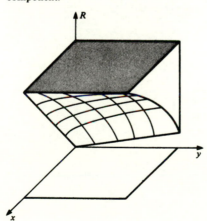

7. Following the text example, calculate numerically the net change in going from $V = p = 1$ to $V = p = 2$ of the quantity whose increment ΔS is given by

$$\Delta S = \frac{c_1}{V} \Delta V + \frac{c_2}{p} \Delta p$$

using routes a, b, c (fig. 16.5). (Use increments $\Delta V = \Delta p = 0.1$ and leave c_1, c_2 arbitrary.)

8. Use formula (16.8) to find the function $S(p, V)$ whose incremental change is given above. Plot the p, V contours of S taking $c_1 = 2.5$ and $c_2 = 1.5$. (*i*) Calculate $S(2, 2) - S(1, 1)$; (*ii*) evaluate $(\partial p / \partial V)_S$ in terms of p, V.

9. Apparatus like that shown in fig. 16.6 is often used as classroom demonstrations of the equilibrium of three forces (see chapter 4). Draw this arrangement (for three unequal weights) on a sheet of graph paper. Remembering that the potential energy of each weight is given by mgh, where h is the height above some fixed level, work out how to estimate the potential energy of the system for any given position of the knot. By doing this for a number of points, *sketch* the contours of potential energy on the graph paper. Locate the minimum as accurately as time permits. Check this position for consistency with the 'triangle of forces' rule (chapter 4).

16.8. Distribution of the intensity of infra-red radiation in a region of the sky. (Courtesy of B. McBreen, D.T. Jaffe and G.G. Fazio.)

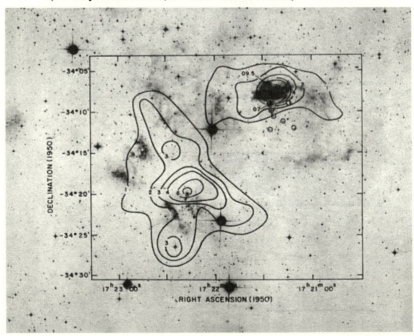

10. Figure 16.7 shows a graphical representation of a function of two variables – in this case, the reflected intensity R from an optical component, as a function of two design parameters. Draw a sketch of the corresponding contours in the (x, y) plane.

11. Figure 16.8 shows the result of some astronomical observations of infrared radiation. Locate the points at which the radiation is a (local) maximum. What is the maximum value over the entire field of view? Over what fraction of the rectangular area does the radiation intensity exceed 50% of this value?

17

Integration

When flat car batteries are first put on the charger the ammeter registers maximum current, say six amps. After an hour or so the current has perceptibly fallen, and if left charging all day (ten hours), eventually sinks to practically zero. Typically the readings at hourly intervals might look as follows:

Elapsed time t (hours)	0	1	2	3	4	5	6	7	8	9	10
$f(t)$, current (amps)	6.0	4.9	4.0	3.3	2.7	2.2	1.8	1.5	1.2	1.0	0.8

for a constant charging rate there would be no difficulty in calculating the accumulated charge, and even for the declining one above an estimate is easy enough. After all, ten hours at an average charging rate of about $(6.0 + 0.8)/2$ amps gives 34 amp-hours. But for greater accuracy a little more labour is required. We could add the first nine values of current (as if each applied as a constant for one hour) – this gives 28.6 amp-hours, but is clearly an over-estimate. What we really need is the *definite integral of the continuous function* $f(t)$ defined by

$$\int_0^{10} f(t)\mathrm{d}t = \lim_{\Delta t \to 0} \sum_{n=0}^{N-1} f(n\Delta t)\Delta t \tag{17.1}$$

where $N\,\Delta t = 10$.

To help us think more about this, we can make a simple plot of the above data as in fig. 17.1. Our first rough estimate replaced the curve by a straight line and then found the area of the triangle bounded by it and the two axes. The area could be obtained more accurately by counting the mm^2 squares beneath the curve taken between zero and ten hours. The errors arising at the curved boundary could be reduced by not counting squares which are estimated to contribute areas less than 0.5 mm^2 and counting 1.0 for those which contribute areas more than 0.5 mm^2. Errors of overcounting and undercounting roughly cancel. An estimate by this method gave 26.0 amp-hours.

The mathematical process of obtaining an accumulated sum from a rate function, numerically or by other means, is called integration. The old name was quadrature, referring to a square or other area. As the present example shows, the units are those appropriate to the product of the rate dimension with the interval dimension – in this case, amps × hours, which could also be expressed in units of charge (Coulombs) with a suitable conversion factor.

There is an alternative view of the integration process. From fig. 17.1 the accumulated amp-hours after successive hours of charging time may be counted and tabulated as follows:

Elapsed time (hours)	1	2	3	4	5	6	7	8	9	10
$F(t)$, charge (amp-hours)	5.4	9.9	13.6	16.6	19.0	21.0	22.7	24.0	25.1	26.0

These points are plotted as the function $F(t)$ shown in fig. 17.2. The relation between $F(t)$ and $f(t)$, (fig. 17.1) is now quite clear. Since $F(t)$ is the

17.1. Charging current as a function of time.

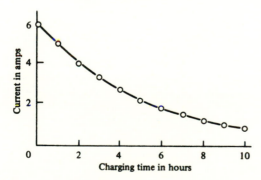

accumulated amp-hours at time t, the difference $F(t + \Delta t) - F(t)$ is the increase in amp-hours in the interval Δt, i.e. the rate $f(t)$ multiplied by Δt.

Thus

$$\frac{F(t + \Delta t) - F(t)}{\Delta t} = f(t).$$

In the limit $\Delta t \to 0$ the left-hand expression becomes the derivative of F, so that $F'(t) = \mathrm{d}F/\mathrm{d}t = f(t)$. The integration of $f(t)$ is therefore the same process as that of calculating a function $F(t)$ whose derivative is $f(t)$. Integration is the inverse of differentiation. (This is the 'Fundamental Theorem of Calculus'.) Graphically, $f(t)$ in fig. 17.1 is just a plot of the slope of $F(t)$ in fig. 17.2.

The solution of the equation $\mathrm{d}\phi/\mathrm{d}t = f(t)$ is called the *indefinite integral* of f and is expressed in a formal way as

$$\phi(t) = \int f(t)\,\mathrm{d}t, \quad \text{or sometimes} \quad \int \mathrm{d}t\, f(t).$$

The symbol $\int$ is a fancy S standing for 'sum', since the process of integration can be expressed in terms of a summation, as above.

17.2. Accumulated amp-hours as a function of time.

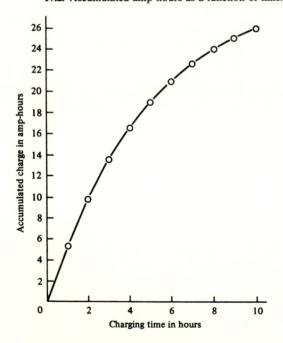

Accumulated charge in amp-hours

Charging time in hours

Quite generally, solutions (or integrals) of the equation $d\phi/dt = f(t)$ are only fully determined when the value of $\phi(t)$ is known at some given instant, because any constant C can be added and the sum remains a solution, since $dC/dt = 0$. The *indefinite* integral must contain an *arbitrary constant*, C. In our example this corresponds to the charge held by the battery before the charging process began. Only the *added* charge is given by the definite integral between t_1 and t_2. That is,

$$\int_{t_1}^{t_2} f(t)\, dt = \phi(t_2) - \phi(t_1). \tag{17.2}$$

Note how the arbitrary constant cancels out on the right-hand side so that it does not affect the value of the definite integral.

If instead we choose to deal with the total charge on a battery, including the pre-existing charge, then we must know the latter (or have some other piece of information) in order to precisely specify the charge at a later time.

These general rules for integration become particularly useful when the rate function $f(t)$ can be expressed analytically, i.e. in terms of known functions, and a corresponding expression for the indefinite integral can be found. In the illustration of battery charging, a simple mathematical model which would adequately describe the electrolytic cell's behaviour leads to the expression $f(t) = I_0 \exp(-kt)$, where I_0 and k are given constants. The dimensions of I_0, k would be those of current and inverse time respectively. Thus, measuring t in hours, k would be so much per hour, i.e. its unit is hour^{-1}. The indefinite integral $\phi(t)$ then satisfies the equation

$$\frac{d\phi}{dt} = I_0 \exp(-kt), \tag{17.3}$$

that is to say, a function $\phi(t)$ is sought whose derivative is proportional to an exponential function. From chapter 13 it is known that $(d/dx)\exp x = \exp x$ and, using the 'function of a function' rule (chapter 13),

$$\frac{d}{dt}\exp(-kt) = -k\exp(-kt). \tag{17.4}$$

Multiplying up by $-I_0/k$ (which does not affect the differentiation), a comparison shows that the indefinite integral must be $\phi(t) = -I_0 k^{-1}\exp(-kt) + C$.

The data presented in fig. 17.1 can be fitted by taking $I_0 = 6$ amps, and $k = 5$ hours^{-1}. As a check, at $t = 5$ hours $f(t) = 6 \times \exp(-5/5) = 6 \times \exp(-1) = 2.2$. Thus the charge accumulated between $t = 0$ and $t = 10$ is given by the definite integral

$$\int_{t_1}^{t_2} I_0 \exp(-kt)\mathrm{d}t = -\frac{I_0}{k}[\exp(-kt_2) - \exp(-kt_1)]$$

where $I_0 = 6$ amps, $1/k = 5$ hours, $t_2 = 10$ hours, $t_1 = 0$.
  Thus the result is

$$6 \times 5 \times (1 - \exp(-10/5)) = 25.9 \text{ amp-hours.}$$

  The constant C disappears in the difference $F(t_2) - F(t_1)$ and is not explicitly needed for this calculation. But to fit the function plotted in fig. 17.2, which vanishes at $t = 0$, the constant C must satisfy

$$-I_0/k \exp(0) + C = 0,$$

i.e. $C = I_0/k$. Thus the graph actually represents the function

$$F(t) = \frac{I_0}{k}[1 - \exp(-kt)] \qquad (17.5)$$

where, as explained, $I_0 = 6$ amps, and $k^{-1} = 5$ hours.
  The art of finding solutions to the equation $\mathrm{d}\phi/\mathrm{d}x = f(x)$ for specified functions of x is cultivated to an agonising degree by teachers and pupils of mathematics. It is the inverse process of differentiation, but whereas there are rules for the latter which will always lead straightforwardly to an answer, algebraic integration is possible only for a limited variety of functions $f(x)$. It is easy to differentiate, say, $x^{-1}\sin x$ but impossible to integrate it in a 'closed form' (though certain related definite integrals are known). The elementary method generally is to reduce the integral to a known standard form such as $\exp(kx)$, $\tan x$, or $(1 + x)^{-1}$. This is achieved by changing the variable of integration. As in the case of differentiation, we shall not list any formulae for integrals of functions here, but will state the rules which are most useful in practice.
  (i) *Change of variables* by the substitution $u = f(x)$. Remember to change the limits, the integrand and the differential $(\mathrm{d}u = f'(x)\mathrm{d}x)$ in the integral.
  (ii) *Integration by parts.*

$$\int u \, \mathrm{d}v = uv - \int v \, \mathrm{d}u. \qquad (17.6)$$

  While analytic methods work more straightforwardly for differentiation than for integration, the reverse is true when numerical methods are employed. Let us take a closer look at the numerical approach, which we adopted rather casually at the beginning of the chapter, when confronted with numerical data.

We could achieve an immediate improvement in the simple numerical integration that we made, if we could use the values of current at the midpoint of each of the time intervals. This might be written

$$\int_{t_1}^{t_2} f(t)\,dt \approx \Delta t \sum_{n=1}^{N} f(t_1 + n\Delta t - \tfrac{1}{2}\Delta t) \qquad (17.7)$$

where each value is taken at the midpoint of an interval of length

$$\Delta t = N^{-1}(t_2 - t_1). \qquad (17.8)$$

This is called the Midpoint Rule. With a little thought, you can see that the error is now $O(\Delta t^2)$, while our earlier estimate (without the $\tfrac{1}{2}\Delta t$) had an error of $O(\Delta t)$. [Multiply the error associated with each interval by the number of intervals required for a given range.] However, we were given the information regarding the charging rate in a form not appropriate to this rule i.e. it was specified 'on the hour' rather than 'on the half-hour'. In this case a somewhat similar rule is more appropriate. This is the Trapezoidal Rule

$$\int_{t_1}^{t_2} f(t)\,dt \approx \Delta t \left[\tfrac{1}{2}(f(t_1) + f(t_2)) + \sum_{n=1}^{N-1} f(t_1 + n\,\Delta t) \right] \qquad (17.9)$$

where Δt, N and $t_2 - t_1$ are related as before. It would be exact if the charging rate varied linearly within each interval, and the error is again $O(\Delta t^2)$. We leave it as an exercise for the student to repeat the numerical

17.3. Estimate of the integral $\int_1^5 \ln x\,dx$ using the Midpoint Rule with N points, as a function of N. The dashed line corresponds to $5\ln 5 - 4$.

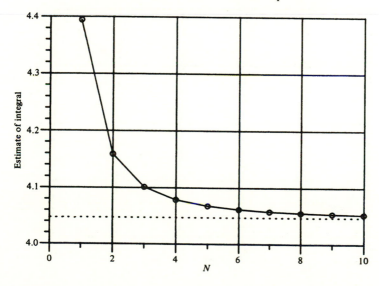

calculation for the battery with this method, and compare with the previous estimates.

Figures 17.3 and 17.4 further illustrate the above remarks using the integral $I = \int_1^5 \ln x \, dx$.

Further rules can be elaborated, such as Simpson's Rule (error $O(\Delta t^3)$), and the rules of Gaussian integration (using unequal time-steps), but nowadays they are of less significance than when the integrals were worked out with a quill pen. The Midpoint or Trapezoidal Rules or even the direct use of (17.1), with a small enough time interval, will often suffice in a computer calculation. Lastly, there is a method which is quite different in spirit to any of the above – the Monte Carlo method. In this, all of the values $f(t_n)$ which contribute to the estimate are equally weighted, as in the Midpoint method, but the points are chosen randomly between t_1 and t_2. For a sufficiently large number of points this will give an adequate estimate of the integral as

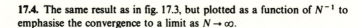

$$\int_{t_1}^{t_2} f(t) \, dt \approx \frac{t_2 - t_1}{N} \sum_{n=1}^{N} f(t_n). \qquad (17.10)$$

Looking back to chapter 2, we see that we have already met an example of this. The calculation shown in fig. 2.4 can be regarded as an estimate of $\int_0^1 t \, dt$, which is exactly $\frac{1}{2}$, by the Monte Carlo method. From the previous discussion, we can expect that the error in this case will be

17.4. The same result as in fig. 17.3, but plotted as a function of N^{-1} to emphasise the convergence to a limit as $N \to \infty$.

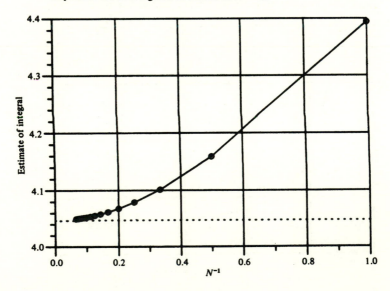

$O(N^{-\frac{1}{2}})$, so it is, at first sight, greatly inferior to our other rules. Figure 17.5 confirms this for the same integral as used before.

The Monte Carlo method is an unnecessarily blunt instrument for such simple integrals – it comes into its own when integrals must be performed over complicated regions in two (or higher) dimensions. For our purposes it is enough to note that it works at all. Note that, as its name implies, it is somewhat of a gamble. In principle, one might get a very unusual sequence of random numbers, giving a wildly inaccurate estimate for the integral. If enough numbers are taken, this becomes overwhelmingly unlikely.

One might think that integration, like a series expansion, must stop whenever a singularity is encountered, at which the function itself diverges. This is not necessarily the case – in certain cases one can integrate up to (or right through!) such a singularity. Such 'improper' and 'principal value' integrals go a little beyond what we wish to do at this stage, but exercise 10 gives an example.

Summary

The integration process is

(*i*) the accumulation of a varying quantity whose rate of change is a known function;

(*ii*) the inverse process of differentiation.

17.5. Monte Carlo estimate of the same integral as in fig. 17.3, as a function of the number of points used.

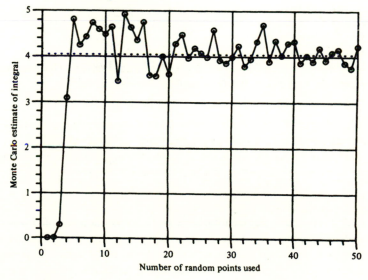

The indefinite integral of $f(x)$ (including an arbitrary constant) is the equation $d\phi/dx = f(x)$, and is written $\int f(x)dx$. The definite integral of $f(x)$ is written $\int_{x_1}^{x_2} f(x)dx$ and given by $\phi(x_2) - \phi(x_1)$.

EXERCISES

1. Any elementary numerical integration of the data given for battery charging will suffer from the inaccuracy of the data itself. Taking this at ± 0.05 amps, estimate the resulting uncertainty in the estimate of the integral.

2. Integrate numerically the experimental curve shown in fig. 13.1 of chapter 13. (Use the same units as the text.)

3. Integrate numerically (by curve plotting and square counting) the function $f(x) = 1/x$ between $x = 1.0$ and $x = 5.0$. Use mm^2 paper with $\Delta x = 0.1$ represented by 1 mm. Plot the result against x.

4. Evaluate the definite integral
$$\int_0^\infty \exp(-ku)du, \quad \text{where} \quad \int_0^\infty du \text{ means } \lim_{x \to \infty} \int_0^x du$$

5. The work done by a thermally insulated gas when it expands from volume V_1 to V_2 is given by $\int_{V_1}^{V_2} p\,dV$ where the pressure p is related to V through the law $pV^{1.4} = $ constant. Evaluate the work done when the initial p, V values are 10^6 Nm^{-2}, 1.0 litre and the final volume is 2.5 litres. (A graphical or algebraic method is acceptable.)

6. Evaluate numerically the definite integral
$$\int_1^2 (2x - x^2)^{\frac{1}{2}} dx.$$
Use the iterative formula
$F(x + \Delta x) = F(x) + (2x - x^2)^{\frac{1}{2}} \Delta x$
from $x = 1.0$ to $x = 2.0$, taking $\Delta x = 0.025$, and $F(0) = 0$. The required integral is then $F(2.0)$. Relate this to the methods in the text.

7. Use the same method as exercise 6 to evaluate $\int_0^1 (\cosh x)^{-1} dx$.

8. Most of our discussion of integrals has centred upon the replacement of an integral by a sum for purposes of definition or calculation. Here is an example of the reverse procedure, which is sometimes useful in examining the convergence properties of series. Consider the approximation
$$\ln n! = \sum_1^n \ln n \approx \int_{\frac{1}{2}}^{n+\frac{1}{2}} \ln x\,dx. \tag{17.11}$$
Give a supporting argument for such an approximation, together with a sketch. While the left-hand side cannot be summed in any simple way, the integral is easy. Show that it gives a result similar to Stirling's approximation (exercise 7 of chapter 14) and identical to it, apart from terms which increase more slowly with n, as $n \to \infty$.

9. Show that
$$\int_0^{\pi/2} \sin^2 x \, dx = \frac{\pi}{4}$$
by relating $\sin^2 x$ to $\cos 2x$ or to $\cos^2 x$. The result is worth remembering in the form – the *average value* of $\sin^2 x$ or $\cos^2 x$ is $\frac{1}{2}$. Of course this only applies over a suitable range; 0 to π or any range which spans an integer multiple of π will do. This average arises frequently in AC circuit theory, in calculating the mean square value of a quantity (e.g. current) which varies sinusoidally.

10. In integrating the function $(x^2 - 1)^{-1}$ from 0 to $a > 1$, the singularity at $x = 1$ can be avoided by defining the Principal Value as the limit as (positive) $\varepsilon \to 0$ of
$$\int_{1+\varepsilon}^a \frac{dx}{x^2 - 1} + \int_0^{1-\varepsilon} \frac{dx}{x^2 - 1}.$$
Evaluate these integrals and show that the limit is $\ln[(a - 1)/(a + 1)]$. Investigate the same integral, for $a = 2$, by the Monte Carlo method.

18

The differential equation

Differential equations are the very essence of much of mathematical physics. In the nineteenth century there was a tendency to replace them with equivalent 'variational principles'. They have since reasserted themselves as the most convenient mathematical expression of many physical laws, at least in their application. Let us see how they arise in some simple examples.

An object, initially at temperature T_0, is heated in the open air. How does its temperature T vary? That is, what is the function $T(t)$? As usual, we must first give a little thought to the physics of this situation. It may be assumed that there is a fixed ratio between the gain (or loss) of heat and the rise (or fall) of temperature, the ratio being the 'heat capacity' C of the object. Given this, the problem is attacked simply by identifying the mechanisms of heat gain/loss. Suppose heat is supplied by the heater at the rate $Q(t)$ (not necessarily constant). From this we must subtract the rate of heat loss to the surrounding atmosphere, which clearly depends on the temperature T of the object – the hotter it gets, the greater the loss. A common assumption is the simple proportionality of this rate of heat loss and the excess of T over the temperature T_a of the surroundings.

Now, if U is the total of the heat transferred to the object after time t, its derivative is the rate of transfer, so

$$\frac{dU}{dt} = Q(t) - \kappa(T - T_a)$$

where κ (Greek kappa) is a constant. Using the assumption of constant heat capacity this becomes

$$C\frac{\mathrm{d}T}{\mathrm{d}t} = Q(t) - \kappa(T - T_a). \tag{18.1}$$

This completes the formulation of the mathematical problem, in the form of a *differential equation* for $T(t)$. Because T and T' enter the equation linearly (no powers, square roots etc.) it is classified as a *linear* differential equation. Note that t is not involved in this definition.

Various physical quantities are necessary to specify the detailed form of the above equation. Even when these are fully defined, we shall see that its solution $T(t)$ would not be fully determined without the additional information, already given, that $T = T_0$ at $t = 0$.

In further exploring the equation, let us assume that no heat is supplied, $Q(t) = 0$ for all t, so that

$$C\frac{\mathrm{d}T}{\mathrm{d}t} + \kappa T = \kappa T_a. \tag{18.2}$$

We still have an interesting problem – that of an object cooling from a temperature above that of its surroundings, of $T_0 > T_a$. (When, as in this case, T' depends only on T and not on t the equation may be called autonomous.)

A feeling for the nature of the solution is best acquired by attempting a numerical solution, as discussed in the next chapter. Formal treatments usually give a 'general solution', containing an arbitrary constant somewhere within it.

As a second example, in which the independent variable is a position rather than a time interval, consider the variation of atmospheric pressure with height. In the simplest model the pressure $P(h)$ depends only upon the height h above sea-level, and horizontal changes are ignored. Then a simple physical argument requires that for a column of air the pressure (force per m^2) at $h + \Delta h$, namely $P(h + \Delta h)$, differs from the pressure $P(h)$ by the weight (in Newtons, say) of a column of air of horizontal cross-section $1\,\mathrm{m}^2$ and thickness Δh. If the (local) density of air is ρ (in $\mathrm{kg\,m}^{-3}$, and h-dependent), then the weight (or force) per m^2 is simply $\rho g \Delta h$, where g is acceleration due to gravity.

From the static balance of forces,

$$P(h + \Delta h) = P(h) - \rho g \Delta h.$$

To complete the relation it is only necessary to relate the density ρ to pressure P. For example, under isothermal conditions (same temperature everywhere) and assuming the perfect gas law, we may use $\rho = \alpha P$ (Boyle's law) where the constant α has a fixed value for air at some temperature. Thus in the limit of infinitesimal changes, the equation for P can be expressed in differential form

$$\frac{dP}{dh} + \alpha g P = 0. \tag{18.3}$$

We have arrived at much the same equation as previously considered, in this quite different context. In both cases, there is a unique relation between the unknown function and its first-order derivative. Here again, the equation only governs the variation of P with height – we need, say, $P(0)$ if we are to complete a solution.

Differential equations can also express a rather different kind of relationship, involving probabilities, as in the following example.

From casual and sometimes frustrating experience, it might seem that certain bus companies operate their services on a statistical basis. To quantify this, suppose that the probability per second of the arrival of a bus is constant and equal to T^{-1}, where T is a time interval fixed by the company. The function $p(t)$ is now introduced to denote the probability that after an interval t (commencing at any instant), no bus has arrived. Then using the rule that independent probabilities can be multiplied to give the resultant composite probability, it must follow that

$$p(t + \Delta t) = p(t)(1 - \Delta t / T).$$

The first factor is the probability that no bus arrive up to time t, while the second is the probability that no bus arrive in the succeeding interval Δt either. In the limit of infinitesimal Δt, the relation can be re-arranged to give the first-order differential equation for $p(t)$, namely

$$\frac{dp}{dt} + \frac{1}{T} p = 0. \tag{18.4}$$

Again, the mathematical resemblance to equations (18.2) and (18.3) should be apparent. In this case, it is clear that $p(0) = 1$, and this is sufficient to determine the solution.

An example which does not explicitly involve time arises from the mass/velocity relation in rocketry. The burnt fuel of the rocket engine expands at high speed to provide reactive thrust which accelerates the rocket. The balance of action and reaction is expressed mathematically through the conservation of momentum in the rocket direction. An increase in the rocket momentum mv is balanced by the creation of exhaust gas momentum in the opposite direction. If a constant (relative) exhaust speed

V is assumed then this momentum is $V\Delta m$, where Δm is the increment of m (actually negative since m is *decreasing*). Thus conservation of momentum becomes

$$m\Delta v + V\Delta m = 0.$$

Proceeding once again to the infinitesimal limit, the differential equation becomes

$$\frac{\mathrm{d}m}{\mathrm{d}v} + \frac{1}{V}m = 0 \tag{18.5}$$

The mathematical form of this equation is identical to that of (18.3) and (18.4). Its solution provides the changing mass m as a function of rocket speed v, or equivalently, v as a function of m. Only if the fuel burning rate $m(t)$ is known independently is it possible to study time-development. In that event it might also be necessary to modify (18.4) to include a gravitational force. As with the earlier differential equations, the complete solution of (18.5) requires an initial value of m, say m_0, corresponding to $v = 0$.

A simple electric circuit (fig. 18.1) provides the final example. Here we wish to find the current $I(t)$, possibly in the presence of a driving voltage $V(t)$. We can write down the total change in potential around the circuit by combining Ohm's law and the law of inductance. Thus a changing current $I(t)$ flowing in an inductance L is opposed by an emf $L\mathrm{d}I/\mathrm{d}t$, while the further emf RI is required to drive it through a series resistance R. If the total external driving voltage is $V(t)$ then the emf balance directly gives the differential equation

$$L\frac{\mathrm{d}I}{\mathrm{d}t} + RI = V(t). \tag{18.6}$$

Linear differential equations like (18.2) and (18.6), which include a term independent of I and I', are called *inhomogeneous*. You should distinguish clearly between 'autonomous' and 'homogeneous'.

Not all first-order differential equations are of the type considered here;

18.1. Electrical circuit made up of an inductance L and a resistance R.

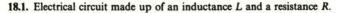

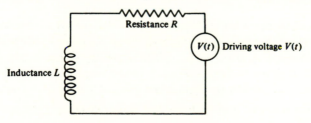

they may involve powers of the first derivative, second derivatives, or even higher. Most important is the extension to the real space of three dimensions (requiring *partial* differential equations). A more general type of causal relation depends upon past history and requires the development of integral rather than differential equations.

The theory and solution of some first-order equations is considered in chapter 19. Before we embark on it, note that the solution of at least one special kind of differential equation is already familiar to us at this point. This is

$$\frac{dy}{dx} = f(x) \tag{18.7}$$

from which we can recover (at least formally)

$$y = \int f(x)\, dx.$$

Remember that this will contain an arbitrary 'constant of integration', which is the expected arbitrary constant. However, in general the variable y will crop up on the *right-hand* side in (18.7), and then the equation cannot be unravelled so easily.

Also, it may already be self-evident that (18.3), (18.4) and (18.5) are equivalent to (10.1), so that the solution of each must be an exponential function. This will indeed emerge in the next chapter.

Summary

A differential equation is an equation which involves a function $y(x)$ and its derivatives, as well as the independent variable x.

A first-order differential equation involves only the first derivative dy/dx.

In a linear equation, y and its derivatives appear in a linear combination.

An initial condition or other supplementary information is necessary to fully define the solution – otherwise it will contain an arbitrary constant.

An autonomous differential equation involves only y and its derivatives.

EXERCISES

1. A well-insulated tank is filled with hot water initially at temperature T_h. Any water drawn off is replaced with an equal amount of cold water at the fixed temperature T_c. Obtain a differential equation relating the temperature T to the fraction x of water drawn off. (The temperature of the mixture is given by the appropriate average of those of its constituents.)

2. Water is poured steadily into a tank at the rate of $0.1\ m^3$ per minute. The

area of the bottom of the tank is 1 m^2. A leak at the bottom causes water to run out at the rate $0.2\,h^{\frac{1}{2}}\text{m}^3$ per minute, where h is the depth of water in metres. (This is known as Torricelli's Law.) Set up a differential equation for h. Without solving, describe the expected form of the solution.

3. The law of mass action in chemistry states that the velocity of a chemical reaction (A + B ⇌ AB) is proportional to the product of the concentrations of the reactants. Show that this implies an equation of the form
$$\dot{y} = k(a - y)(b - y)$$
for the concentration of the product of a reaction.

4. A raindrop (assumed to be spherical) grows in the atmosphere at a rate λA, proportional to its surface area A. Formulate an equation for its radius as a function of time.

5. The *partial* pressure of one of the constituents of a mixture of gases is given approximately by the ideal gas law $pV = nRT$, where n is the number of moles of that constituent. Partial pressures add to give total pressure. A sample of water and dry air is placed in an enclosure which is maintained at constant temperature and pressure. The water evaporates at a rate proportional to the difference between the partial pressure of H_2O in the air and the saturated vapour pressure, which is fixed by T. What form of differential equation is obeyed by $x(t)$, the partial pressure of H_2O? Draw a *rough* sketch of the expected behaviour of $x(t)$.

6. The flow of water into a tank is controlled by a tap which is coupled to the water level and closes as the level is raised. Set up a simple first order differential equation relating the water level h, its maximum h_0, and the flow rate.

7. A car axle is vertically constrained by the suspension spring force $-kz$ together with the shock absorber force $-a\dot{z}$, where z is its vertical displacement and a, k are positive parameters. Write down the force balance equation for an arbitrary force $F(t)$. Discuss the solution $z(t)$ in the case of a jolting force $f(t)$ of large magnitude but short duration.

8. Use a sheet of graph paper to *sketch* the solutions of exercise 3 as follows, for $k = a = b = 1$. Choose a grid of points and at each point draw a small segment of slope equal to the value of y' at that point. Join up the segments to sketch a family of curves.

19

Solving first-order differential equations

Mathematicians, nothing if not logical, have a prior question before searching for solutions to differential equations – do such solutions exist? Before that even – what constitutes existence? For the pure mathematician satisfactory answers to such questions constitute the theory of differential equations. The physicist generally adopts a pragmatic attitude to this. Apart from questions relating to initial or other conditions on the solution, it is generally obvious that the equation has a solution. If we impose too many or too few conditions on the solutions, we shall find no solution or one that is not unique, and must think again.

Let us begin with the *numerical solution* of the kind of differential equation which occurred repeatedly in the previous chapter. Suppose that when numerical values of physical quantities are substituted, we find that we wish to obtain a solution of

$$\frac{\mathrm{d}T}{\mathrm{d}t} + 2T = 10 \tag{19.1}$$

for t up to $t = 3$, given that $T = 20$ at $t = 0$.

We shall use a *step-by-step* method, in which the required time range is divided up into equal intervals, which are processed successively. It is closely connected to the elementary methods of integration discussed in chapter 17 – indeed integration is just a special case of what we are doing here, as we remarked at the end of the last chapter.

Suppose we choose to use 15 time intervals, each of length 0.2 units. We use these like stepping stones across a river, as an alternative to swimming continuously across it. How do we get from the first one, $t = 0$, to the second, $t = 0.2$?

The differential equation (together with the initial conditions) tells us that the derivative dT/dt at $t = 0$ has the value -30. Approximating T by a linear function, in the way now familiar to us, we can write

$$T(\Delta t) \approx 20 + \Delta t \, \dot{T}(0). \tag{19.2}$$

Note the usefulness of the dot (or prime) notation for derivatives in this kind of equation – it makes it possible to display the point at which it is evaluated in a neat manner. Use this to estimate T at the next point

$$T(0.2) \approx 20 + 0.2 \times (-30) = 14. \tag{19.3}$$

If we accept this estimate for $T(0.2)$, we have thus achieved one (approximate) step. The problem posed by the next one is precisely the same. We estimate dT/dt at $t = 0.2$, again using the differential equation, and hence

$$T(0.4) \approx 14 + 0.2 \times (-18) = 10.4. \tag{19.4}$$

In this way we continue until we have covered the required range. Figure 19.1 shows the result, compared with the exact solution which is derived later in this chapter.

19.1. Results of the Euler method, applied to the equation $dT/dt + 2T = 10$ with a steplength $\Delta t = 0.2$, compared with the exact solution $5\{1 + 3 \exp{(-2t)}\}$ (dotted line).

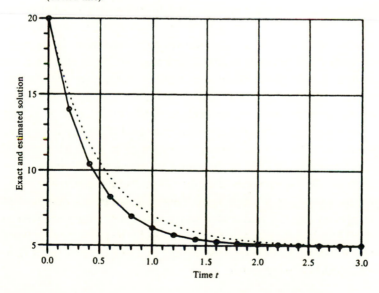

Clearly this method (the *Euler method*) applies equally well to any equation of the form

$$\frac{dy}{dx} = f(x, y). \tag{19.5}$$

It does not matter if the derivative is a function of both variables y and x, nor how complicated it is, provided that it can be evaluated for given x and y.

Note also that we required to specify a starting temperature. If we change the value of this, we generate a different solution. There is a family of possible solutions, which we could distinguish by the starting temperature. Usually a real physical problem will contain some piece of information to pick out the required solution from this family. It could be the starting temperature (an *initial condition*, or more generally a *boundary condition*) or something more subtle – for example, we might require to find the time at which dT/dt takes some particular value. This would equally well pick out a particular member of the family of solutions.

Euler's method is really very crude – it has an error $O(\Delta x^2)$ per step where Δx is the steplength. It is an easy matter to improve upon this, notably by the *improved Euler method,* which achieves an error $O(\Delta x^3)$ with a little more computation. The defining equations of this for equation (19.5) are

$$\tilde{y} = y(x) + \Delta x f(x, y(x)) \tag{19.6}$$

$$y(x + \Delta x) = y(x) + \Delta x \times \tfrac{1}{2}(f(x, y(x)) + f(x + \Delta x, \tilde{y})). \tag{19.7}$$

The first equation is the same as that used to estimate $y(x + \Delta x)$ in the elementary Euler method but here it gives only a provisional value, leading to the improved estimate (19.7). There are yet more refined methods in the same spirit, which are generally available as library routines. However, the computational power of the modern computer makes even the crudest methods quite effective in many cases. Roughly speaking, the more sophisticated methods 'think harder' before finally taking each step, and can take correspondingly longer steps, with increased efficiency. If we carry this approach to its logical conclusion, we might sit at $t = 0$, generate the sequence of derivatives $d^n T/dt^n$ at that point by repeated differentiation of (19.5), construct the Taylor series (chapter 14) up to high order, and make the transition to $t = 3$ in a single leap! This is possible in principle and indeed in practice for the case under study, but a warning was given in chapter 2 of the problems that may be encountered when series methods are pushed too far. In practice the improved Euler method, or some other step-by-step procedure, is a better bet.

The search for solutions by analytical methods is not quite as elementary as this and, in describing it, we shall concentrate on the special case in which the equation is linear with constant coefficients, as in chapter 18,

$$\frac{dy}{dx} + ky = F(x). \tag{19.8}$$

Thus in (18.2) (assuming C to be independent of temperature), k is the ratio κ/C, while in (18.6) it is R/L. Since k plays a key role in the solution, it is important to appreciate its physical significance. In all the examples k was positive, and physically this is commonly the case. For the formal solutions of (19.8) k is unrestricted.

The inhomogeneous term $F(x)$ must have the same dimensions as ky. In (18.2) it is simply the constant value $\kappa T_a/C$. Since a constant value (including zero) is common, this special case will be assumed for the moment, and $F(x)$ replaced by F_0. A solution is now sought to the simpler equation

$$\frac{dy}{dx} + ky = F_0. \tag{19.9}$$

Many school mathematicians recall the inversion trick which, essentially, is to regard x as a function of y, and write

$$\frac{dx}{dy} = (F_0 - ky)^{-1}$$

which after an elementary integration gives $x = B - k^{-1}\ln(F_0 - ky)$, where B is the usual 'constant of integration'. After the necessary re-inversion, y is obtained as

$$y = \frac{F_0}{k} - \frac{1}{k}\exp kB\exp(-kx). \tag{19.10}$$

While this procedure gives a formally correct solution it obscures some valuable features of (19.9). Moreover, it suggests (erroneously) that whatever the value of B the coefficient of $\exp(-kx)$ has the opposite sign to k. Worse than this, the mathematical argument is often badly presented and misunderstood.

A preferable and physically intuitive argument goes as follows. The simplest solution to (19.9) is $y = F_0/k$ for which $dy/dx = 0$. In application to (18.2) this becomes $T = T_a$, which represents the final temperature attained when the cooling has slowed down to a negligible level, i.e. after a very long time or, formally, as $t \to \infty$. This solution then takes care of the inhomogeneous term F_0; what remains to be found is a solution to the homogeneous equation, namely $dy/dx = -ky$, to be combined with it. But it is one definition of an exponential function that differentiation just reproduces the function (see chapter 10), i.e. $d/dx \exp x = \exp x$. The factor $-k$ is obtained by replacing x with $-kx$. Furthermore, this function can be multiplied by any factor A whatsoever, positive or negative, to give the general solution, namely $y = A \exp(-kx)$. Finally the two solutions can be added (because y, dy/dx occur linearly) to give

$$y = \frac{F_0}{k} + A \exp(-kx). \tag{19.11}$$

This is formally identical with the earlier solution, taking $A = -k^{-1} \exp(kB)$.

The parameter A is the 'constant of integration' for (19.9). Its continuous range of values provides the family of solutions. Schematically they appear as in fig. 19.2 (drawn for $k > 0$).

Returning to the more general equation (19.8), there is a standard method of obtaining the most general solution. This involves changing the dependent variable $y(x)$ to the new function $Y(x)$, where $Y = y \exp(kx)$. Differentiating the product,

$$\frac{dY}{dx} = \exp(kx) \left[\frac{dy}{dx} + ky \right] = F(x) \exp(kx). \tag{19.12}$$

$Y(x)$ is now obtained by integration as $Y = \int F(x) \exp(kx) dx$, which in turn gives $y = \exp(-kx) Y(x)$. The factor $\exp(kx)$ is described as an *integrating factor* and the method (which can be extended to allow for an x-dependence of k) is described by that name.

In time-dependent problems such as the circuit equation (18.6), the inhomogeneous term might well be sinusoidal, say $V_0 \cos \omega t$, where the

19.2. Family of solutions to a first-order differential equation, $dy/dx = -ky$.

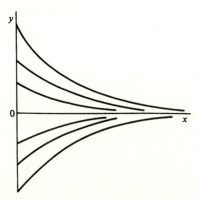

peak voltage V_0, and circular frequency ω are given constants. The integrating-factor method could be applied but a more intuitive method would be as follows. Since the first time derivative of $\cos \omega t$ is $\sin \omega t$ it is reasonable to assume that a solution for $I(t)$ can be constructed as a linear combination of the two functions, i.e. $I = a \cos \omega t - b \sin \omega t$. Substitution of this form into (18.6) and equating coefficients of $\cos \omega t$ and $\sin \omega t$ gives a pair of simultaneous equations for a and b, namely

$$Ra - \omega Lb = V_0$$
$$\omega La + Rb = 0$$

Solving,

$$a = RV_0[(\omega L)^2 + R^2]^{-1}$$
$$b = -\omega LV_0[(\omega L)^2 + R^2]^{-1} \tag{19.13}$$

This solution contains no constant of integration and in fact is analogous to the simple solution $y = F_0/k$ for (19.9). The most general solution is obtained by adding the term $A \exp[-(R/L)t]$ which solves the corresponding homogeneous equation. Such a contribution (with arbitrary constant A) represents a *transient* or decaying current, as opposed to the oscillatory behaviour of the other term in the solution.

Summary

First-order differential equations may be numerically integrated by elementary step-by-step methods.

The first-order linear differential equation of the form

$$\frac{dy}{dx} + ky = F_0$$

has as its most general solution

$$y = \frac{F_0}{k} + A \exp(-kx)$$

where A is the arbitrary constant.
When F_0 is replaced by $F(x)$, the integrating factor is $\exp(kx)$.

EXERCISES

1. Use the improved Euler method, specified in the text, to obtain a solution of (19.1) with the same range and step length as in fig. 19.1 and compare its accuracy with that of the elementary Euler method.

2. In retrospect, how might the numerical calculation of chapter 10 be described in terms of a differential equation and a method of solution?

3. A tank contains 100 l of brine in which 10 kg of salt is dissolved. Fresh water runs in at the rate of 5 l/min. and the mixture (well stirred) runs out at the same rate. Formulate and solve a differential equation for the amount of salt in the tank as a function of time.

4. A projectile of mass 1 kg is fired vertically with a propelling force proportional to elapsed time, equal to αt, where $\alpha = 2 \, \text{Ns}^{-1}$ (allowing for gravity). It is subject to a drag force proportional to velocity, given by βv where $\beta = 0.1 \, \text{Nm}^{-1} \text{s}$. Describe its motion, and find its position after ten seconds.

5. Complete the solution of exercise 2 of chapter 18.

6. *Non-linear* equations of the same type as

$$\frac{dy}{dt} = y(1 - y)$$

for a population $y(t)$ are popular with ecologists and economists. They provide some relief from the Malthusian doctrine of inevitable exponential increase to disaster! Sketch and discuss the expected form of the solution (a) for $y(0) = 0$, (b) for $y(0) = 1$, (c) for $y(0) = 0.1$. For the last case, give a numerical or analytical solution leading to an estimate of $y(3)$. Can you think of a model for, say, population growth which would suggest such an equation? Also give a further discussion of exercises 3 and 8 of chapter 18.

7. Complete the solution of exercise 4, chapter 18.

20
Second-order differential equations

If a particle of mass m is subject to a force which is directed towards a point O and proportional to the distance from O, it can execute simple harmonic (i.e. sinusoidal) motion along a straight line through 0, its point of equilibrium. This is the familiar motion of a mass which hangs on a spring, and all vibratory motion of mechanical systems, however complicated, can be mathematically related back to this simple case.

The equation which dictates the form of this motion is a second-order differential equation. It is Newton's second law, *force = mass times acceleration*, which gives us

$$\text{acceleration} = \text{mass}^{-1} \times \text{force},$$
$$d^2x/dt^2 = -m^{-1}\lambda x. \tag{20.1}$$

Here x is distance along the straight line, m is mass and λ is the 'Hooke's Law constant' which relates force to displacement. The reader may find the change of notation from the last chapter unsettling – but in physical science one must learn to recognise the essence of the equations, regardless of the variables used. Let us look at some examples of the occurrence of this type of equation, which we shall call the *oscillator* equation.

Consider a particle at rest in the bottom of a bowl. If displaced from the centre and released, it oscillates about its original point of equilibrium. This motion is due to the restoring force of gravity together with the reaction force (fig. 20.1). The specification of these forces will lead us to the oscillator equation, given certain approximations. The standard treatment of the problem involves resolving vectors and using Newton's second law but it is really quite subtle. Instead we shall adopt an easier but somewhat back-handed approach and use the principle of conservation of energy:

potential energy + kinetic energy = constant

that is,

$$mgf(x) + \tfrac{1}{2}m\dot{x}^2 + \tfrac{1}{2}m\dot{z}^2 = \text{constant} \tag{20.2}$$

where $f(x)$ is the height of the bowl as a function of radius. This assumes that motion is confined to one (x, z) plane. (For a more general discussion, see chapter 27).

Two approximations are now applied, both depending on restriction of the motion to small displacements from equilibrium. We expand $f(x)$ in a Taylor series and keep only the first significant term, i.e.

$$f(x) = cx^2 \tag{20.3}$$

Note that because we are expanding about the equilibrium point (a stationary point) there is no linear term in x. Secondly, on the same assumption, the small contribution to the kinetic energy from vertical motion is neglected. These approximations are not guesswork; they may be shown to correctly describe the motion in the limit of small displacements. Now we have

$$mgcx^2 + \tfrac{1}{2}m\dot{x}^2 = \text{constant}. \tag{20.4}$$

The constant depends on the particular motion which is set up – we can get rid of it by differentiating with respect to time, obtaining

$$\ddot{x} + 2cgx = 0 \tag{20.5}$$

or, using $\omega^2 = 2cg$ for later convenience

$$\ddot{x} = -\omega^2 x \tag{20.6}$$

which is indeed the oscillator equation. Comparing this with Newton's second law, we identify the right-hand side as the restoring force (divided by mass).

20.1. Particle sliding in a bowl.

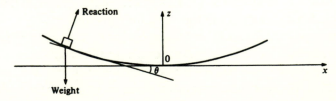

The whole argument applies equally well to the familiar case of the simple pendulum (fig. 20.2). Here the function f is given by the appropriate equation for an arc of a circle of radius l,

$$f(x) = l - (l^2 - x^2)^{\frac{1}{2}}. \tag{20.7}$$

Hence, from the series expansion of $f(x)$ (chapter 15) $c = (2l)^{-1}$ and $\omega^2 = gl^{-1}$ for the pendulum.

Clearly the same equation applies when *any* mechanical system is slightly disturbed, provided that only one variable (here, x) is involved. If there is more than one variable ('more than one degree of freedom'), we are led to the use of matrices (chapter 9) in sorting out the possible simple harmonic motions of the system, to each of which (20.6) separately applies.

An example of great concern to engineers is the elastic deformation of a material. By applying suitable external forces, all solid bodies can be strained away from their natural shape. Provided the force is not too large, in springy or elastic materials it produces a strain y (fractional change in dimensions) which is proportional to the force. In the strained equilibrium position this force is exactly balanced by a resisting internal force which must therefore be proportional to $-y$. If now the body is released from the outside force, then the unbalanced internal force accelerates each part of the body at the rate proportional to $\ddot{y}$. It follows that the motion is mathematically described by an equation $\ddot{y} = -\omega^2 y$, where the coefficient ω^2 depends upon the 'springiness' or elasticity of the material, and also its density.

However, we have so far left out an important and sometimes dominant physical effect, namely *friction*. Mechanical systems do not go on freely vibrating for ever. Frictional effects always 'damp', i.e. reduce the amplitude of, the motion, eventually making the system settle down at equilibrium unless some further force is applied.

20.2. Simple pendulum.

In general, forces which oppose motion and dissipate energy associated with it may be called *dissipative* – they include the familiar force of friction and the drag or viscous force associated with motion in a fluid. Since the force always opposes the motion, it is often taken as $-\gamma \, dx/dt$, where γ (Greek gamma) is positive. In real applications this may be only a first approximation, but the addition of further terms, e.g. higher powers of dx/dt, will not change the general behaviour of the solutions very much. Mechanical friction between solid objects is constant in magnitude, which is actually a little more awkward mathematically. The force proposed here best describes the viscous drag of a fluid, such as the air through which the pendulum moves. With the incorporation of this, (20.6) becomes

$$m\frac{d^2x}{dt^2} + \gamma\frac{dx}{dt} + m\omega^2 x = 0. \tag{20.8}$$

In this more general form, the coefficients of $\ddot{x}$, $\dot{x}$, and x are all constant. Because of this, the equation is exactly soluble in terms of the functions which we have studied.

One further extension of (20.8) is to include an external *driving force F(t)* in place of the zero on the right-hand side. There is an interesting and useful parallel in the theory of linear electric circuits. A circuit containing an inductance L in series with a resistance R was considered in chapter 18. Suppose to this circuit is added (in series) a capacitor of capacitance C (see fig. 20.3). Then the voltage required to produce electric charge Q, $-Q$ on the plates is Q/C. In charging or discharging this capacitor a current dQ/dt flows. Since charge cannot accumulate elsewhere, $dQ/dt = I$ (current in circuit). Then expressing dI/dt as d^2Q/dt^2, the total opposing emf of the cell becomes $L \, d^2Q/dt^2 + R \, dQ/dt + Q/C$, so that balancing this with a driving

20.3. Electrical circuit made up of an inductance L, resistance R and a capacitance C.

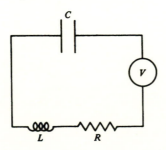

voltage $V(t)$ gives the *circuit equation*:

$$L\frac{d^2Q}{dt^2} + R\frac{dQ}{dt} + \frac{Q}{C} = V(t).\tag{20.9}$$

The presence of $V(t)$ on the right hand side of (20.9) means that the equation is inhomogeneous. It is possible to excite the circuit first with $V(t)$ and then switch off the driving voltage, i.e. put $V(t) = 0$ thereafter. The resulting homogeneous equation is then entirely similar to (20.8).

It is worth comparing the roles of the constant coefficients. Inductance L corresponds to mass m, resistance R to drag coefficient γ. But note that the force coefficient (of x) corresponds to C^{-1}, not C, that is to say, small electrical capacitance corresponds to large mechanical restoring force (for given displacement).

In practice, $V(t)$ would often be sinusoidal, say $V(t) = V_0 \cos\Omega t$. The solution corresponding to that of the L, R circuit (chapter 18) is given in chapter 23.

Before we embark on a detailed analysis of this very promising equation let us decide in advance what we expect, using commonsense, experience or physical ideas.

If the added drag (or resistance) term is small we must expect to see the same oscillations as are obtained without the term, only slightly modified. Such intuitive ideas of 'continuity' are used throughout physics and only occasionally come unstuck. The effect of the small term in the mechanical system is always to oppose the motion and hence the system does work against it and loses energy. The oscillations must die away.

If, on the other hand, the viscous force is very large, the situation is rather different. Consider the fate of a pendulum in porridge. We expect the system to creep back to equilibrium very slowly, without oscillation.

Engineers sometimes design systems, such as dashboard instruments, to be 'critically damped' so they come to equilibrium as fast as possible without oscillating. This is the borderline case between the two we have suggested.

The solution of the circuit equation is obviously worth examining in some detail. Since a mathematical diversion will be involved in doing so, it will take three more chapters to achieve this goal.

Summary

The oscillator equation $\ddot{x} = -\omega^2 x$ gives a very general description of oscillation about equilibrium but it is often necessary to add damping, by means of a term proportional to $\dot{x}$. The resultant second-order linear differential equation with constant coefficients describes

(*i*) mechanical oscillation, generally appearing in the form

$$m\ddot{x} + \gamma\dot{x} + m\omega^2 x = 0;$$

(*ii*) electrical L, R, C circuits, appearing as

$$L\ddot{Q} + R\dot{Q} + C^{-1}Q = 0.$$

A driving force or voltage may be added to such an equation, making it inhomogeneous.

EXERCISES

1. Find the linear second-order differential equation satisfied by $y = A\exp(-x) + B\exp(-2x)$, independently of A, B. (Eliminate A, B from y, dy/dx, d^2y/dx^2.)

2. By the same method as in exercise 1, find the second-order differential equation of which $x = A\exp(-3t)\cos(4x + B)$ is a solution for any A and B.

3. What is the consequence of changing the sign of γ in the equation (20.8)? Can you envisage any cases where this might be appropriate?

4. Reverse the order of the discussion given in the text, deriving the energy conservation law (20.2) from Newton's second law in the form (20.5).

5. Resistance in a circuit plays a role analogous to friction or viscous drag in a mechanical one. Comment on the underlying physics of this, including their relevance to consideration of energy. Can you think of any further examples of physical systems to which the circuit equation might apply?

21

Solving second-order differential equations

In this chapter we shall first reflect upon the procedure for solving second-order differential equations numerically, just as we did with first-order equations. Indeed, the obvious method is closely similar. For simplicity, we shall focus on the oscillator equation (20.6). Again we shall temporarily disregard the fact that our chosen equation has a simple analytic solution. This is 'simple harmonic motion', $x = A \sin(\omega t + B)$ or $C \sin \omega t + D \cos \omega t$. Instead we shall ask how we would set about solving it numerically as there are some general lessons to be learned from this. Let us try to find the analogue for second-order equations of Euler's method already studied for first-order equations. Again we shall choose convenient numerical values for purposes of illustration, so we take $\omega^2 = 4$ and will look for a solution in the range $t = 0$ to $t = 4$.

We shall try to find this by a step-by-step procedure, with a steplength $\Delta t = 0.1$. But again we need a 'jumping-off point' – some starting information. If we are given, say, $x = 1$ at $t = 0$ we can use the differential equation to get d^2x/dt^2 at $t = 0$, but this is not enough information to make the required step. We need to be given (or choose) a starting value for dx/dt, say 2, as well, and use *two* equations to get started. Using the dot notation for derivatives, these are

$$x(0.1) \approx x(0) + 0.1\,\dot{x}(0) = 1 + 0.1 \times 2 = 1.2 \tag{21.1}$$

$$\dot{x}(0.1) \approx \dot{x}(0) + 0.1\,\ddot{x}(0) = 2 - 0.1 \times 4 = 1.6 \tag{21.2}$$

The first is our familiar linear approximation for x, the second is the same, applied to $\dot{x}$. The differential equation is used only to work out the term which is underlined.

Just as in the previous case, this procedure can be used on each successive step until we have reached the end of the specified range. This is Euler's method for a second-order differential equation. The error per step is $O(\Delta t^3)$.

Note that we already know that the solution of this problem oscillates with a period $2\pi/(4)^{\frac{1}{2}} = \pi$. It would make no sense at all to try to solve it numerically by our method, using a steplength as large as this; the linear approximations involved would be grossly in error. Often one can guess at such a limitation in a more complicated problem, by comparison with a simpler one like this. Indeed, it is always necessary to give some thought to this question of the maximum reasonable steplength before proceeding.

In figs. 21.1–3 solutions are shown which illustrate these remarks. The choice $\Delta t = 0.1$ for steplength gives reasonable results, but $\Delta t = 0.01$ gives much closer agreement with the exact solution $x(t) = \sin 2t + \cos 2t$. Figure 21.3 shows the disastrous consequences of too large a steplength, $\Delta t = 1$, giving a qualitatively unsatisfactory result.

There is another way of looking at the equations (21.1) and (21.2), and indeed at the differential equation itself. If we give the first derivative dx/dt the same status as x itself, any linear second-order equation can be rewritten

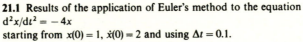

21.1 Results of the application of Euler's method to the equation $d^2x/dt^2 = -4x$ starting from $x(0) = 1$, $\dot{x}(0) = 2$ and using $\Delta t = 0.1$.

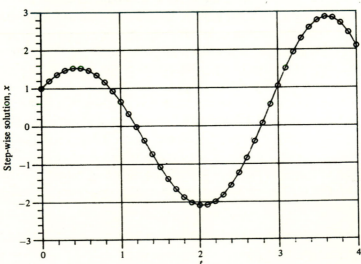

21.2. Same as fig. 21.1, using smaller steplength $\Delta t = 0.01$ (individual points not shown).

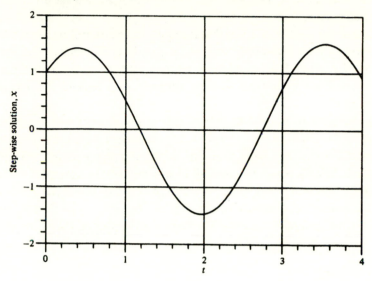

21.3. Illustration of the effect of much too large a steplength. The same method and equation were used as in fig. 21.1, but with $\Delta t = 1.0$.

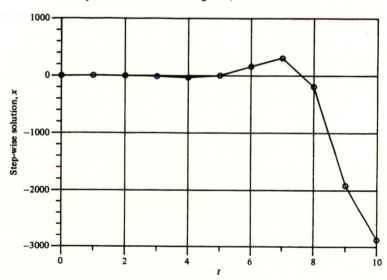

as two linear first-order equations, to be solved simultaneously for the two functions x and dx/dt. Computer library programs are often set up in this way. (see also exercise 5.)

To summarise, we have learned that it requires two conditions to specify the solution of a second-order differential equation. Formally, this will mean that its general solution must contain two arbitrary constants. (*Second* order/*two* conditions/*two* parameters; this should be easily remembered.)

Our investigation into the analytical side of the subject will again be restricted to the linear case (i.e. linear in the dependent variable x, its first and second derivatives). Note that this is not restricted to the case of constant coefficients which includes all of our immediate examples.

Recall the general form which we found for the linear first-order differential equation (summary of chapter 19). By analogy with this, we may expect the general solution of a linear second-order equation for $x(t)$

$$a(t)\, d^2x/dt^2 + b(t)\, dx/dt + c(t) = d(t) \tag{21.3}$$

to take the form

$$x = Af(t) + Bg(t) + h(t) \tag{21.4}$$

where A and B are arbitrary constants.

We shall not prove this rigorously, but note its plausibility. The functions $f(t)$ and $g(t)$ are two solutions of the homogeneous equation

$$a(t)\frac{d^2x}{dt^2} + b(t)\frac{dx}{dt} + c(t) = 0 \tag{21.5}$$

obtained by setting to zero the right-hand side of (21.3). (These two solutions might be obtained, for example, by integrating (21.5) with two different sets of initial conditions.) Clearly any linear combination of the two, i.e. $Af(t) + Bg(t)$, together with any particular solution $h(t)$ of the full equation, is still a solution. Substitute (21.4) in (21.3) and note how the f and g terms drop out.

It is useful to remember the geometric significance of d^2x/dt^2 as the *upward curvature* of the function $x(t)$. Thus, in our first example in this chapter the solution curves downwards when it is above the axis $x = 0$, upwards when it is beneath it. This is why the function wiggles up and down about the axis in an oscillatory manner. On the other hand, if d^2x/dt^2 had the same sign as x, as in the equation $d^2x/dt^2 = 4x$, the curvature is always away from the axis, and the solution cannot be oscillatory. In this particular case it is *exponential* and the word may be used loosely for this general behaviour in other cases.

A given equation may have solutions whose character is oscillatory and exponential in different ranges. Atomic wave functions, which are solutions of a linear second-order differential equation, have this feature (fig. 21.4).

At this point, we still lack a specific analytical form for the solution of the general equation with constant coefficients. The derivation of this requires an addition to our mathematical armoury – the theory of complex numbers. This is introduced in the next chapter.

Summary

Second-order differential equations may be solved by step-by-step methods analogous to those used for first-order equations. Both the solution function and its derivative are estimated at each step.

Linear second-order differential equations have solutions which take the form

A, B arbitrary constants

$$Af + Bg \qquad + h$$

| Solutions of the homogeneous equation | Any solution of the inhomogeneous equation |

It is often useful to classify solutions qualitatively as being *oscillatory* or *exponential* in some region.

21.4. Typical atomic wave function.

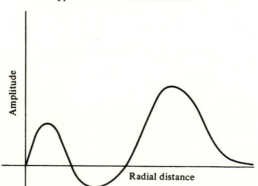

EXERCISES

1. Give the general solution of $d^2x/dt^2 = 4$ and show how this conforms to the theory of linear second-order equations.

2. A mass of 0.1 kg oscillates with a displacement x (in metres) satisfying the equation $\ddot{x} + 100x = 0$. Find the solution which corresponds to any energy of 0.1 J and an initial ($t = 0$) speed of $5.0\,\text{ms}^{-1}$.

3. Write down the most general solution to the equation

   $$\frac{d^2y}{dx^2} = k^2y \quad (k^2 = \text{constant}).$$

   Find the most general solutions which satisfy each of the following conditions:

   (i) $y(-x) = -y(x)$

   (ii) $y'(0) = 0$

   (iii) $\displaystyle\int_0^\infty y^2\,dx = k^{-1}$

4. Solve the differential equation

   $$\frac{d^2y}{dx^2} + \frac{dy}{dx} + y = 0$$

   numerically, with the initial conditions $x = 0$; $y = 1$, $dy/dx = 0$. Plot y against x between $x = 0$, $x = 6$. Compare with the exact solution. (Assume $y = F \exp mx$ and solve for m to find this.)

5. Explain how the idea of representing a second-order equation as *two* first order equations would be implemented for

   $$\frac{d^2y}{dx^2} = -y$$

   by writing equation for y and dy/dx. Relate these to the methods used in chapter 11.

22

The complex exponential

For the practical physicist and engineer complex numbers provide a handy tool for the analysis of AC circuits. A few basic mathematical rules enable reactive circuit elements like inductances and capacitances to be treated with nothing more difficult than Ohm's law. The first-order linear equation (18.6), dealing with the voltage/current relationship in a simple *LR* circuit can be used to illustrate the benefits of the complex approach. We shall really just be repeating the steps which led us to (19.13) but we shall see them in a different light.

For generality it is helpful to regard the driving voltage as having both cos and sin components, expressed as $A \cos \omega t - B \sin \omega t$ (the choice of sign is to comply with later conventions). Then the responding current also has both components and, following chapter 19, is written $a \cos \omega t - b \sin \omega t$. The mathematical object is then to calculate the coefficients a, b in terms of A, B by equating the driving voltage to the voltage fall across the elements L, R, as before.

Now the voltage change across the resistance is simply proportional to the current so that the cosine and sine oscillations of the voltage are the same, multiplied by R. In engineering terms this is the 'quadrature response'. On the other hand the inductance requires a time differentiation which effectively exchanges the components (with a change of sign) to give the voltage fall $-\omega L a \sin \omega t - \omega L b \cos \omega t$. In a 2×2 matrix notation (chapter 7), the voltage balance equations (obtained by comparing coefficients of cos ωt, sin ωt) become

$$\begin{bmatrix} A \\ B \end{bmatrix} = \left[R\begin{pmatrix} 1 & 0 \\ 0 & 1 \end{pmatrix} + \omega L \begin{pmatrix} 0 & -1 \\ 1 & 0 \end{pmatrix} \right] \begin{bmatrix} a \\ b \end{bmatrix} \tag{22.1}$$

On the right-hand side, R is accompanied by the 2×2 unit matrix I. The inductive reactance L is associated with the matrix $\begin{pmatrix} 0 & -1 \\ 1 & 0 \end{pmatrix}$ which rotates the vector (a, b) anticlockwise through an angle of $90°$. The equation may be written and manipulated more simply by replacing the matrix by the symbol i (engineers often prefer j), and rewriting the above equation as

$$V = (R + i\omega L)I. \tag{22.2}$$

Here the two-component vectors V, I are now considered to constitute *complex numbers*. In the case of the current we can write

$$\begin{bmatrix} a \\ b \end{bmatrix} = \left[a + \begin{pmatrix} 0 & -1 \\ 1 & 0 \end{pmatrix} b \right] \begin{bmatrix} 1 \\ 0 \end{bmatrix} \leftrightarrow I = a + ib,$$

the unit vector $\begin{pmatrix} 1 \\ 0 \end{pmatrix}$ being dropped. Evidently from its matrix definition, $i^2 = -1$. In terms of the effect of the matrix, multiplying any vector, this is equivalent to saying that two anticlockwise rotations of $90°$ give rotation of $180°$, which *reverses* the vector.

In this and in many cases, the complex notation is just a way of avoiding having to deal with 2×2 matrices and vectors. But complex numbers have much more to offer than this.

The introduction of a quantity i, defined to satisfy the equation

$$i^2 = -1, \tag{22.3}$$

enabled the development of the important mathematical field of 'complex analysis'. In this i is described as 'imaginary'. Then, as above, the quantity $a + ib$ (with a, b real) is called a complex number, consisting of real and imaginary parts. With practice such numbers can be manipulated by the usual rules of algebra, replacing i^2 by -1, i^3 by $-i$ etc. wherever such multiples crop up. Two additional pieces of notation must be remembered. If $z = a + ib$, its modulus $|z|$ is given by $|z|^2 = a^2 + b^2$, while its complex conjugate $\bar{z}$ (sometimes written z^*) is simply $\bar{z} = a - ib$.

The theory of complex numbers makes a neat connection between the cosine and sine functions and the exponential function. According to the definition and calculation of chapter 11, the basic iterative equations for cos x and sin x may be written in 2×2 matrix form

$$\begin{bmatrix} \cos(x + \Delta x) \\ \sin(x + \Delta x) \end{bmatrix} = \begin{bmatrix} \cos x - \Delta x \sin x \\ \sin x + \Delta x \cos x \end{bmatrix}$$

$$= \left[I + \begin{pmatrix} 0 & -1 \\ 1 & 0 \end{pmatrix} \Delta x \right] \begin{bmatrix} \cos x \\ \sin x \end{bmatrix}$$

$$(22.4)$$

In complex notation the 2×2 multiplying matrix becomes $1 + i\Delta x$, while the accompanying vector becomes $\cos x + i \sin x$. It follows that this latter complex function (call it $f(x)$) satisfies the equation

$$f(x + \Delta x) = (1 + i\Delta x)f(x).$$

By comparison with the exponential definition (chapter 10), and noting that $f(0) = 1$, it follows that

$$f(x) = \cos x + i \sin x = \exp ix. \qquad (22.5)$$

The great Swiss mathematician Leonard Euler (1707–83) was responsible for this synthesis of exponential and trigonometric functions, and much else – he was the most prolific contributor to the development of mathematics in Europe. The enigmatic special case $\exp(i\pi) = -1$ is often quoted as encapsulating a couple of hundred years of mathematical development.

The series expansions which we examined in chapter 14 provide a further demonstration of Euler's Theorem since, from the series definition of the exponential function,

$$\exp ix = 1 + ix + \tfrac{1}{2}(ix)^2 + \dots. \qquad (22.6)$$

Separation into real and imaginary parts provides the series expansions of $\cos x$ and $\sin x$ (equations (15.9, 15.10)). This should also set at rest any fears regarding the meaning of $\exp(ix)$.

The practical use of Euler's Theorem always involves *taking real and imaginary parts*, so let us study this operation. For example, its application to $(x + iy)^{-1}$ is achieved by so-called *rationalisation*, namely

$$\frac{1}{x + iy} = \frac{x - iy}{(x + iy)(x - iy)} = \frac{x}{x^2 + y^2} - \frac{iy}{x^2 + y^2}. \qquad (22.7)$$

The real and imaginary parts are then clearly $x(x^2 + y^2)^{-1}$ and $-iy(x^2 + y^2)^{-1}$. Note that the factor i is sometimes omitted, the imaginary part being quoted as simply $-y(x^2 + y^2)^{-1}$.

These ideas already enable us to complete the solution of the *LR* circuit problem in a compact form.

Solving the *complex* equation (22.2) for *complex I*,

$$I = (R + i\omega L)^{-1} V \tag{22.8}$$

we find the required solution by taking real and imaginary parts. For example, if $A = V_0$ and $B = 0$, rationalisation gives

$$a = \frac{V_0 R}{R^2 + \omega^2 L^2}, \quad b = -\frac{V_0 \omega L}{R^2 + \omega^2 L^2}. \tag{22.9}$$

This is the same solution (19.13) that we found previously. The mathematical acrobatics which we have performed could hardly be justified by this alone, but chapter 23 will offer at least a hint of the power of such a method.

The Cartesian coordinates provide a further method of representing a complex number – as a point in a plane, with coordinates representing real and imaginary parts. This kind of graphical representation is called the *Argand diagram*. It is easily seen (fig. 22.1) that the complex number $I \exp(i\omega t)$ moves in a circle about the origin with radius

$$|I| = (a^2 + b^2)^{\frac{1}{2}} = V_0 (R^2 + \omega^2 L^2)^{-\frac{1}{2}}.$$

To most of us it seems strange that the rather arbitrary generalisation which introduces complex numbers should prove so fruitful in practical mathematics. Hamilton, who first used the above compact notation, carried the generalisation even further to define 'quaternions' with four components. These never found the same breadth of application as the complex

22.1. Unit circle $|z|^2 = 1$ in the Argand diagram.

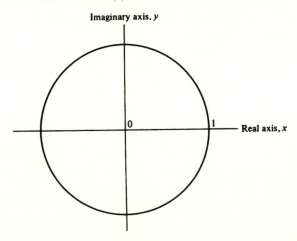

Imaginary axis, y

0   1   Real axis, x

numbers. They played an important part in the liberation of mathematics from the traditional rules of algebra which are appropriate to real numbers.

Summary

Complex quantities are algebraic expressions $a + ib$ (real part + i × imaginary part), where i satisfies the equation $i^2 = -1$. If the real and imaginary parts are regarded as two components of a vector, the symbol i may be represented by the 2×2 matrix $\begin{pmatrix} 0 & -1 \\ 1 & 0 \end{pmatrix}$.

The modulus of $a + ib$ is $(a^2 + b^2)^{\frac{1}{2}}$. The complex conjugate is $a - ib$. The complex exponential $\exp(ix)$ is identical to $\cos x + i \sin x$ and can be used to analyse sinusoidal behaviour in AC circuit theory.

EXERCISES

1. If a and b are complex numbers show that (*i*) $|ab| = |a||b|$, (*ii*) $|a + b|^2 = |a|^2 + |b|^2 + 2\,Re(a*b)$.
   Apply these rules to evaluate the complex impedance z,
   $z = [(i\omega L + R)^{-1} + i\omega C]^{-1}$, giving $|z|$ and phase of z.

2. Find all solutions of the equation $z^4 = -\frac{1}{2} + i\sqrt{3}/2$ by writing z in the form $V \exp i\theta$ where V and θ are real.

3. Evaluate the real part of $\exp(mx)$, where $m^2 + m + 1 = 0$. What second-order differential equation does $A \exp(mx) + B \exp(m*x)$ satisfy?

4. In the discussion of the first-order equation (19.9), two arbitrary constants arose which were said to be equivalent, being related by $A = k^{-1} \exp(kB)$. If $A = -1$ and $k = 2$, what is B?

5. Use the identity $(\exp ix)^3 = \exp 3ix$ to show that $\cos 3x = 4\cos^3 x - 3\cos x$.

6. Evaluate the imaginary part of $\exp(i\omega t)\,[\omega - \omega_0 + i\gamma]^{-1}$.

7. Show that
   $$\frac{1}{1 + \exp(ix)} = \tfrac{1}{2} + \tfrac{1}{2}i\,\tan\frac{x}{2}$$
   Discuss its behaviour as (*i*) $x \to \pi$, (*ii*) $x \to \pm i\infty$.

8. A well-known definite integral is $\int_0^\infty \exp(-\alpha t)\,dt = 1/\alpha$, provided $\alpha > 0$. Assuming this identity to hold even if α is given an imaginary component, say $\alpha \to \alpha + i\beta$, use the result to obtain values for
   $$\int_0^\infty \exp(-\alpha t)\cos\beta t\,dt \quad \text{and} \quad \int_0^\infty \exp(-\alpha t)\sin\beta t\,dt.$$

9. Plot the real and imaginary parts of the complex solution to $qz = 1 + z^2$ as functions of real q for $-\infty < q < +\infty$.

10. A particle of charge e and mass m moving in a uniform magnetic field $\mathbf{B}$ has equations of motion (see exercise 10, chapter 5)
$$\dot{v}_x = \omega_c v_y, \quad \dot{v}_y = -\omega_c v_x,$$
where $\omega_c = eB/m$, and v_x, v_y are the components of velocity transverse to the magnetic field $\mathbf{B} = (0, 0, B)$. By considering $v_x + iv_y$ show that solutions can be expressed as $v_x = Re\, V\exp(-i\omega_c t)$, $v_y = Im\, V\exp(-i\omega_c t)$ where V is an arbitrary complex velocity. Explain the physical significance of the phase and modulus of V.

23
The circuit equation

Let us try out the use of complex exponentials in the oscillator equation, which in the anonymous $y(x)$ notation is

$$\frac{\mathrm{d}^2 y}{\mathrm{d}x^2} + \omega^2 y = 0. \tag{23.1}$$

The general solution to this was recognised early in chapter 20, as a linear combination of $\sin \omega x$ and $\cos \omega x$, in the present notation. We must expect to recover this, whatever way we treat the equation. For present purposes, we substitute the trial solution $y = \exp(\lambda x)$ and observe that this gives solutions for $\lambda = \pm i\omega$. Hence the general solution is

$$y = c_1 \exp(-i\omega x) + c_2 \exp(i\omega x) \tag{23.2}$$

where c_1 and c_2 are (complex) constants. This is, in general, complex and we ultimately want a real solution, so we insist that its imaginary part be zero. This leads immediately to the requirement that c_1 and c_2 be complex conjugates, say $c_1 = a + ib$ and $c_2 = a - ib$. Then, gathering up the real part of (23.2), it can be written (using equation (22.5) as

$$y = 2a \cos \omega x + 2b \sin \omega x, \tag{23.3}$$

which is the expected solution. Remember that a and b are arbitrary constants so the factors of two have no significance.

Now we can attempt a similar, and more interesting, analysis of the circuit equation, (20.9) whose solution is not so familiar or self-evident.

Recall that its general solution is the sum of the so-called 'particular integral', which has no arbitrary constants) and the 'complementary function', which is the general solution of the corresponding homogeneous equation. The physical significance is that the complementary function describes transient currents which flow as a result of switching on and off. The particular integral describes the current which follows (though not in phase) the driving voltage, after the transient currents have died away.

The complementary function satisfies the equation

$$L\frac{dI}{dt} + RI + \frac{Q}{C} = 0 \tag{23.4}$$

where $I = dQ/dt$. With applications in mind, the notation is left unaltered and possible simplifications of changes of variables are ignored. Again we use a trial solution $I(t)$ of the form $\exp(\lambda t)$, where λ (with the dimensions of inverse time) is a parameter whose possible values are to be determined. Substitution into (23.4) gives an equation for λ:

$$L\lambda + R + 1/C\lambda = 0. \tag{23.5}$$

The last term arises from the relation $I = dQ/dt$ which gives for the assumed form of I, $Q = \lambda^{-1}\exp(\lambda t)$. Solving the quadratic equation (23.5) gives the roots as λ, λ^*, with

$$\lambda = -\frac{R}{2L} + i\left(\frac{1}{LC} - \frac{R^2}{4L^2}\right)^{\frac{1}{2}} \tag{23.6}$$

where it is assumed that $L \geqslant \frac{1}{4}R^2C$. With R of the order of kilohms and C of the order of microfarads this is commonly the case. Thus λ is complex, with a real part which is negative.

The solution to (23.4) is the sum of the complex exponentials $\exp(\lambda t)$, $\exp(\lambda^*t)$ with complex coefficients. As with the simple undamped oscillator, real initial conditions require that the solution be real so that the coefficients must be conjugate. It follows in the same way that $I(t)$ is given by

$$I(t) = (1/2)A \exp i\lambda t + \text{(complex conjugate)}$$

$$= \exp\left(-\frac{R}{2L}t\right)[a\cos\omega t - b\sin\omega t]$$

where

$$\omega = (2L)^{-1}(4L/C - R^2)^{\frac{1}{2}}. \tag{23.7}$$

Putting $a = |A|\cos\phi$, $b = |A|\sin\phi$, this can be written

$$I(t) = A\exp(-R/Lt)\cos(\omega t + \phi).$$

This is a decaying oscillation. Figure (23.1) illustrates such a function. This corresponds to our intuitive guess of the form of the solution for small R, in chapter 21. Note that indeed we have used $R^2 \leqslant 4L/C$. If we instead assume the reverse inequality, then the roots of (23.5) will be real, and the solution will be a sum of two decaying exponentials. Thus $R = 2(L/C)^{\frac{1}{2}}$ is the condition for *critical damping*.

Turning now to the particular integral of (20.9), this must depend on them driving term $V(t)$. We shall take this to be sinusoidal: the voltage $V_0 \cos \Omega t$ to be inserted in the right hand side of (20.9) can be regarded as the real part of a complex exponential. A solution is now sought to the fictitious complex equation

$$L dI/dt + RI + Q/C = V_0 \exp(i\Omega t) \qquad (23.8)$$

where I, Q are necessarily complex. The true (physical) function is then obtained as the real part of the complex current. The imaginary part is not needed, but in any case corresponds to a driving voltage $V_0 \sin \Omega t$ which only differs from the original by the phase $\pi/2$. With a little practice, you will appreciate the virtues of this trick, which at first may seem an unnecessarily indirect approach to finding a real solution of a real equation. It is popular even with practically minded engineers!

Bearing in mind the rule for differentiation of an exponential it is clear that a solution of (23.8) is proportional to $\exp(i\Omega t)$. The calculation then reduces to finding its coefficient, which evidently will be complex. Denoting this by I_0, then

$$I(t) = I_0 \exp(i\Omega t),$$

23.1. Example of a decaying oscillating function.

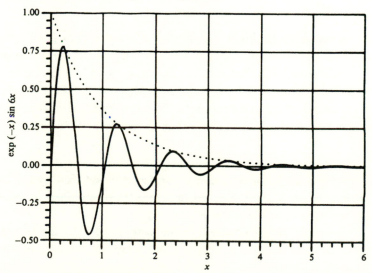

which upon substitution into (23.8) gives

$$(i\Omega L + R + 1/i\Omega C)I_0 = V_0. \tag{23.9}$$

In circuit theory the coefficient of I_0 is known as the *complex impedance* and is simply the sum (for the present *LCR* series circuit) of the imaginary quantities $i\Omega L$, $1/i\Omega C$, and the real resistance R.

Finally, the physical solution is obtained as

Real part of $I(t)$ = Real part of $I_0 \exp(i\Omega t)$.

According to (23.9) this becomes, after slight re-arrangement,

$$I(t) = \text{Real part of} \frac{i\Omega C V_0 \exp(i\Omega t)}{1 - \Omega^2 LC + i\Omega CR}.$$

To deal with the complex denominator, multiply top and bottom by its complex conjugate, $1 - \Omega^2 LC - i\Omega CR$. The real part of the numerator then becomes

$$-V_0(1 - \Omega^2 LC)\Omega C \sin \Omega t + V_0(\Omega C)^2 R \cos \Omega t.$$

The complete expression for the current is given by

$$\frac{(\Omega C)^2 R \cos \Omega t - (1 - \Omega^2 LC)\Omega C \sin \Omega t}{(1 - \Omega^2 LC)^2 + (\Omega CR)^2} \cdot V_0 \tag{23.10}$$

Full discussion of the physical significance of this expression is given in physics texts. It is only necessary to note here that the terms in $\cos \Omega t$, $\sin \Omega t$ indicate a *phase shift* between voltage and current. Thus if the current is expressed as proportional to $\cos(\Omega t - \phi)$, then evidently

$$\tan \phi = \frac{1 - \Omega^2 LC}{\Omega CR}. \tag{23.11}$$

At the frequency Ω_0 for which $1 - \Omega^2 LC = 0$, i.e. $\Omega_0 = (LC)^{-1/2}$, (*i*) the amplitude of the current oscillation is a maximum and (*ii*) ϕ changes sign. For small values of R, the frequency Ω_0 approximates to the value ω given by (23.7). Variation of Ω through the value Ω_0 produces the phenomenon of *resonance*, i.e. a peak in the response of the system when the driving force is given a frequency which matches that of free oscillations.

Summary
The oscillator equation $y'' + \omega^2 y = 0$ is satisfied by the complex functions $\exp(i\omega x)$, $\exp(-i\omega x)$, which may be recombined to recover the two solutions $\cos \omega t$ and $\sin \omega t$.

To solve the circuit equation in the absence of a driving voltage, substitute the trial solution $y = \exp(\lambda t)$ and solve the resulting equation for complex λ. Combine complex exponentials to give real solution of the form $A \exp(-Bt)\cos(Ct + D)$, where A and D are arbitrary. When the damping term is large, exponential solutions are found instead. To solve for a particular integral in the presence of a driving voltage, first replace real voltage $V_0 \cos \Omega t$ by complex $V_0 \exp(i\Omega t)$. Assume solution $I_0 \exp(i\Omega t)$ and calculate complex I_0. Take real part of $I_0 \exp(i\Omega t)$.

EXERCISES

1. The decaying charge on the capacitor for an LCR circuit (see text) is given generally as
   $Q(t) = \exp(-\alpha t)(A \cos \beta t + B \sin \beta t)$
   where $\alpha \pm i\beta$ are the roots of the quadratic equation $L\lambda^2 + R\lambda + 1/C = 0$. Evaluate A, B if it is known that initial $(t = 0)$ charge/current values are Q_0, I_0. How would $Q(t)$ be altered if a constant biasing potential V is included in the circuit?

2. An oscillator of natural frequency ω_0 driven by an external oscillatory force $f \cos \omega t$ has a response $x = f(\omega_0^2 - \omega^2)^{-1} \cos \omega t$. Describe the behaviour at resonance $\omega \to \omega_0$ and relate it to the solution $x = (ft/2\omega_0) \sin \omega_0 t$. Sketch the latter as a function of t.

3. An oscillator of frequency ω_0 is given a sharp impulsive force $F(t)$ described by $F(t) = \text{constant } F_0$ from $-\tau < t < \tau$ and zero otherwise, where $\lim_{\tau \to 0}(2F_0\tau) = P$. Show that the displacement x remains unaltered but that the momentum changes discontinuously by the amount P. How would the response change if a resistive force $\gamma\dot{x}$ were also present?

4. Check that the time-dependent displacement
   $x = f/\omega_0^2 + (a - f/\omega_0^2)\cos \omega_0 t$
   is that of a simple harmonic oscillator subject to a constant biasing force f, with initial $(t = 0)$ displacement $x = a$. At $t = \pi/\omega_0$, the bias is reversed in sign, $f \to -f$. Obtain the value of x at $t = 2\pi/\omega_0$. If this reversing bias cycle is repeated at the same intervals for N times, find the value of x at $t = 2N\pi/\omega_0$, and interpret the result. (See chapter 24 on Fourier series.)

5. Radio waves are reflected and returned to earth by the ionosphere. Regarded as rays they follow curved paths $z(x)$ given by solutions to the differential equation
   $$n^2(0)\cos^2 \alpha \frac{d^2z}{dx^2} = \frac{1}{2}\frac{d}{dz}[n^2(z)].$$
   Here x is the distance from the transmitter at $x = 0$, z is the height above ground, $n(z)$ the variable refractive index, and α is the initial ray angle with the ground,
   Assuming as a model $n^2(z) = 1 - k^2z^2$, obtain a solution for which

$(dz/dx)_0 = \tan \alpha$, and show that the maximum range is $x = \pi k^{-1} \cos \alpha$. Show also that $n(z) \cos \phi$ remains constant over the curve path, where $\tan \phi = dz/dx$. Discuss the relation to Snell's law of refraction.

6. Draw a sketch of the trajectory described by the function
$$f(t) = e^{i\mu t},$$
in the Argand diagram, where μ is complex. How is it related to the solution of the current equation?

7. Supply the proof of the statement, made in the text, that c_1 and c_2 must be complex conjugates in order for the solution (23.2) to be real.

8. Integrate the formula
$$w = \int_0^t \frac{dx}{dt}\left(-\gamma \frac{dx}{dt} \right) dt$$
for a work done against the drag force, for a damped oscillator displaced from equilibrium and released; find the total loss of energy as a function of time. Sketch and discuss the result.

24

Harmonics and Fourier series

Musicians, whether they saw, hammer or blow, are all acquainted with overtones. The octave of a note simply doubles the frequency (or very nearly, for the well-tempered piano) and coincides with its first overtone. An octave plus a (major) fifth trebles the frequency and defines the second overtone, and so on. A single sustained note (unless produced by a good tuning fork) actually contains many such overtones, a fact musically exploited by the harmonies of Claude Debussy. The richness and quality of a musical note depends very much upon its tonal content.

The Greeks, first in the field as so often, developed the arithmetic of tonal or harmonic perfection, incorporating it even into architecture and their ideas concerning the motion of the planets. However, the mathematics of modern harmonic theory is largely due to Joseph Fourier (1768–1830) who used it in his treatise on the propagation of heat. Fourier's idea was to represent a function by a series of sine (and cosine) functions. Although wave motion (with its acoustic applications) and diffusion (with its thermal applications) are mathematically described by different equations (see chapters 25, 26), both may be solved in a very general way by the method of Fourier.

As well as these two areas of physics, Fourier analysis turns up characteristically in the description of electrical signals. As explained in chapter 23, the differential equation which describes the voltage/current relation in an *LCR* circuit is easily solved for a sinusoidal input voltage of arbitrary frequency ω, and shows a resonance at $\Omega_0 = (LC)^{-\frac{1}{2}}$. However in applications to electronic devices neither the driving voltage nor the current need be sinusoidal, though they may still be periodic. An important mathematical problem is presented by the circuit equation for a more general type of periodic voltage drive. Fourier analysis provides a framework for this, and sometimes the only straightforward method of solution. The voltage drive is regarded as being made up of a sine wave of the same period (the *fundamental*), augmented with those of higher frequencies (*harmonics*).

As happens so often, it is the *linearity* of the LCR circuit equation that is the key to its solution. This means simply that if a driving voltage $V_1(t)$ produces a circuit current $I_1(t)$, and a second voltage $V_2(t)$ independently produces a current $I_2(t)$, then the current produced by the combined voltage $V_1(t) + V_2(t)$ is $I_1(t) + I_2(t)$. Such a basic result would be ruined by the inclusion of a non-linear element, say a capacitor whose capacitance depended upon voltage. It follows that if the driving voltage can be regarded as built up from individual sinusoidal components then the resultant current is the sum of the corresponding individual responses, and these are often easily obtained. The key problem then is the resolution or decomposition of a general periodic signal into such sinusoidal components.

Consider a *square wave*, illustrated by fig. 24.1. How are we to find the harmonics which make up such a function? We may start first with the fundamental, i.e. a sine (or cosine) of the same period as the function itself. In this case it is convenient to arrange that the nodal points coincide, as shown in fig. 24.2.

24.1. Square wave (extending in a similar manner between $t = \pm \infty$).

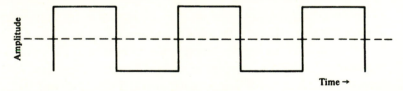

24.2. Square wave and corresponding fundamental.

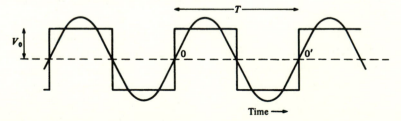

With the origin O taken at such a nodal point, at the centre of the discontinuous rise, the fundamental becomes $B_1 \sin 2\pi t/T$ where T is the basic repeat time (period) of the square wave, i.e. the separation OO'. The choice of phase is arbitrary so that, for example, by relocating O at the centre of a flat section, the sine is replaced by a positive or negative cosine. Next it is necessary to calculate the coefficient B_1, and this is done using the following argument. Denoting the discontinuous square wave function by $V(t)$, then it can be regarded as built up from the fundamental plus an overtone or harmonic remainder. Choosing B_1 correctly means that the remainder, namely $V(t) - B_1 \sin 2\pi t/T$ should not contain an oscillation of period T, but consist solely of *higher* harmonics. Such a function has the useful property that when multiplied by the missing fundamental, the resultant interference produces cancellation and causes the oscillatory product to average to zero,

$$\int_0^T \left(V(t) - B_1 \sin \frac{2\pi t}{T} \right) \sin \frac{2\pi}{T} \, dt = 0. \tag{24.1}$$

Noting the integrals

$$\int_0^T \sin^2 \frac{2\pi t}{T} \, dt = \frac{T}{2},$$

and

$$\int_0^T V(t) \sin \frac{2\pi t}{T} \, dt$$

$$= V_0 \int_0^{T/2} \sin \frac{2\pi t}{T} \, dt - V_0 \int_{T/2}^T \sin \frac{2\pi t}{T} \, dt = \frac{2T}{\pi} V_0$$

it follows that $B_1 = (4/\pi)V_0$. Note however that the area under the positive half-cycle of the fundamental is $(T/\pi)(4V_0/\pi)V_0 \simeq 0.405\,V_0 T$ compared with $0.5V_0$ for the square wave.

The next harmonic must have nodes at $t = 0, T, 2T$, etc., but with additional ones as well. Thus $\sin 4\pi t/T$ is at first a possibility. However, this would give a lop-sided look because (with a positive amplitude) it would increase the left-hand side of the fundamental and decrease the right-hand side. It might seem that $\cos 4\pi t/T$ is better, but this one raises the positive voltage and reduces the negative. So the next possibility is $B_3 \sin 6\pi t/T$. Similar arguments suggest that the series continues with the odd harmonics, with cosine terms excluded.

The amplitudes B_3, B_5 etc., are found by a further application of the cancellation idea. The mathematical principle is that the product of any two different harmonics m, n say, integrated over one cycle, gives zero. Thus using the addition formulae (chapter 11),

$$\int_0^T \sin(2\pi mt/T)\sin(2\pi nt/T)\,dt$$

$$= 1/2 \int_0^T [\cos(2(m-n)\pi t/T) - \cos(2(m+n)\pi t/T)]\,dt$$

$$= 0, \quad \text{for} \quad m \neq n. \tag{24.2}$$

This operation (multiplication followed by integration) may be likened to the taking of a scalar product of two vectors (multiplication followed by summation) (chapter 5). Since their 'scalar product' is zero, the harmonic functions are naturally described as *orthogonal*.

In geometry a vector $\mathbf{v}$ can be resolved into a combination of orthogonal vectors $\mathbf{i}, \mathbf{j}, \mathbf{k}$, and the coefficient of $\mathbf{i}$ is simply the scalar product $\mathbf{i} \cdot \mathbf{v}$. In the same way, if the function $V(t)$ is 'resolved' into its orthogonal harmonics, the coefficient of the m-harmonic is given by the analogous expression $\int_0^T \sin(2\pi mt/T)V(t)dt$. But whereas $\mathbf{i} \cdot \mathbf{i} = 1$ by definition, for the harmonics normalisation is required since $\int_0^T \sin^2(2\pi mt/T)dt = T/2$. It follows then that in the analogy with the corresponding vector formula there is an extra factor arising from this. The result is

$$B_m = 2/T \int_0^T \sin(2\pi mt/T)V(t)\,dt. \tag{24.3}$$

The fundamental coefficient B_1 has already been evaluated, essentially by the same formula ($m = 1$). It has been anticipated from a geometrical argument that $B_2 = 0$; the formula (24.3) confirms this. Quite generally for an odd integer n, $B_n = 4/n\pi$. Figure 24.3 shows successive Fourier approximations to $V(t)$, taking n up to 49.

The important scalar product idea can be extended to include (in general) the cosine terms. It is easy to show that in formula (24.2), either or both of the sine factors can be replaced by a cosine factor with the same argument, i.e. $\sin 2n\pi t/T \rightarrow \cos 2n\pi t/T$ and the integration result remains the same. All this leads to the following basic theory. If a signal $V(t)$ has a repeat interval $t = T$, then $V(t)$ may be expressed as a sum of its cosine and sine harmonics. Thus

$$V(t) = \sum_{m=0}^{\infty} A_m \cos(2\pi mt/T)$$

$$+ \sum_{m=1}^{\infty} B_m \sin(2\pi mt/T). \tag{24.4}$$

The cosine series starts with $m = 0$, i.e. simply the constant A_0. Mathematically the sum over the harmonic order m may be 'to infinity' (with consequent interesting questions of convergence) but physically a signal generator may be unable to produce frequencies $2\pi m/T$ beyond a few megahertz say. This means that the harmonic series is terminated and thus only approximates to the intended signal shape.

Questions arise as to whether both cosine and sine terms should be present. The answers depend upon the symmetry of the signal $V(t)$ with respect to the chosen time-origin. In the square wave signal described by fig. 24.2, $V(-t) = -V(t)$, i.e. it is *odd*. It follows immediately that only sine harmonics are needed, i.e. $A_n = 0$, for all n. If however the time-origin had been taken at the mid-point of a section of the positive voltage then $V(t)$ would have been *even* in t and so composed of cosine harmonics. Furthermore, from fig. 24.2, $V(t + T/2) = -V(t)$. Testing this on the sine terms,

$$\sin[(2\pi n/T)(t + T/2)] = \sin[(2\pi nt/T) + n\pi)]$$

Thus the required behaviour is obtained only if n is odd, i.e. $n = 1, 3, 5, \ldots$ etc. All this just confirms that the square wave signal (with the time-origin as in fig. 24.2) is given by the odd sine harmonic series.

In general, the magnitudes of the harmonic amplitudes are given by (24.3)

24.3. When the Fourier series for the square wave is truncated after a finite number of terms the results are as shown for 1, 3, 5 and 49 terms.

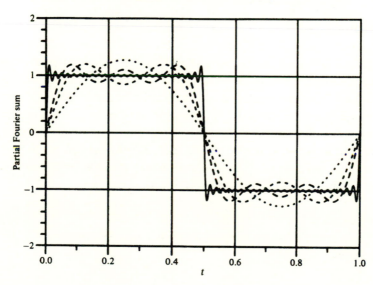

together with the cosine equivalent,

$$A_m = 2/T \int_0^T \cos(2\pi mt/T)V(t)\,dt. \tag{24.5}$$

Note however that (24.5) does not hold for $m = 0$. By simply integrating (24.4) from $t = 0$ to $t = T$ (the sinusoidal integrals vanish) A_0 is equated to the average of $V(t)$, namely

$$A_0 = 1/T \int_0^T V(t)\,dt. \tag{24.6}$$

Returning to the LCR circuit response, the problem can be posed: what form of current (i.e. time-dependence) is produced if a square wave voltage signal of period T is fed into an LCR circuit? To obtain an answer it is first necessary to express the square wave in terms of its Fourier components, namely, writing $\Omega = 2\pi/T$,

$$V(t) = (4V_0/\pi)(\sin \Omega t + \tfrac{1}{3}\sin 3\Omega t + \tfrac{1}{5}\sin 5\Omega t + \ldots), \tag{24.7}$$

and in accordance with principle of linearity, treat each harmonic driving voltage separately. Thus to each harmonic m of the voltage there is a harmonic $I_m(t)$ of the current satisfying the equation

$$L(dI_m/dt) + RI_m + Q_m/C$$
$$= (4V_0/\pi)(1/m)\sin m\Omega t. \tag{24.8}$$

This differential equation can be solved using the complex method of chapter 23, namely, writing $\sin m\Omega t = I_m \exp(im\Omega t)$ and obtaining first the complex response. The total current response is then the sum of all the current harmonics.

In practical cases, it may be necessary to find Fourier coefficients numerically. This boils down to doing integrals (24.3–6) so it poses no great problems, although nowadays there are some particularly efficient ways of doing it (Fast Fourier Transforms, dating from a paper in 1965).

Finally we should note that the Fourier series is often used to represent a function over a finite interval T, and the periodic continuation of this which is represented by the series has no significance. See, for example, exercise 7. The interval T can even be made *infinite*, in which case the series becomes an integral, but we shall content ourselves with series here.

Summary

A function $V(t)$ with a repeat interval T can be represented by the Fourier series,

$$V(t) = \sum_{m=0}^{\infty} A_m \cos(2\pi mt/T)$$

$$+ \sum_{m=1}^{\infty} B_m \sin(2\pi mt/T)$$

where

$$A_0 = \frac{1}{T} \int_0^T V(t)\,dt,$$

$$A_m = \frac{2}{T} \int_0^T \cos(2\pi mt/T)V(t)\,dt$$

$$B_m = \frac{2}{T} \int_0^T \sin(2\pi mt/T)V(t)\,dt.$$

If $V(-t) = V(t)$, then only cosines are present, $B_m = 0$.
If $V(-t) = -V(t)$, then only sines are present, $A_m = 0$.

EXERCISES

1. The text suggests that the operation in (24.2) is like a scalar product. Discuss this (*a*) in terms of algebraic properties (*b*) in terms of an approximation to the integral by a sum (chapter 17).

2. Sketch the function which is defined by the integral of the square wave shown in fig. 24.2. Find its Fourier representation by integrating that of the square wave, term by term.

3. Sketch the function which is the periodic continuation of $f(x) = x$ for $-\pi < x < \pi$. Show that its Fourier series is
$2(\sin x + \frac{1}{2}\sin 2x + \frac{1}{3}\sin 3x + \cdots)$.

4. A faulty signal generator which is set to generate a sine wave voltage of amplitude V_0 instead generates a signal for which voltages $V > V_0/2$ are replaced by $V_0/2$. This is corrected by a filter which removes all higher harmonics. By how much is the final signal reduced in amplitude, relative to V_0?

5. Use the series generated in exercise 2 to show
$$\frac{\pi}{4} = 1 - \frac{1}{3} + \frac{1}{5} - \frac{1}{7} + \cdots.$$

How many terms are needed on the right-hand side to give an error of less than 1%?

6. Combine the results of exercise 2 and exercise 3 to find the Fourier

representation of the function

$$f(x) = 0, \quad -\pi \leqslant x \leqslant 0$$
$$f(x) = x, \quad 0 \leqslant x \leqslant \pi.$$

7. Find the coefficients of the Fourier sine series which represents a function which is defined by

$$y(x,0) = 2ax/L, \quad 0 \leqslant x \leqslant L/2$$
$$= 2a(L-x)/L, \quad L/2 \leqslant x \leqslant L.$$

Give a sketch of the function represented by the series between $-4L$ and $+4L$.

8. A capacitor in series with resistor is charged by a 10 V battery for 1.0 *ms* and then instantaneously discharged by short circuit. The charging is repeated and the cycle continued indefinitely. Assuming a capacitance of $0.1 \mu F$ and a resistance of $1 k\Omega$, calculate the fundamental harmonic coefficient of the charge oscillation. (Note: $\int_0^\infty \exp(-\alpha x) \cos \beta x \, dx = \alpha(\alpha^2 + \beta^2)^{-1}$.)

9. A single atom has an electric charge density whose variation in the x-direction is the even function $\rho_a(x)$, where the atomic centre is at $x = 0$. A one-dimensional atomic chain is formed by placing very many similar atoms at intervals d along the x-direction to give a resultant density

$$\rho(x) = \sum_{n=-\infty}^{\infty} \rho_a(x - nd).$$ Show that the mth Fourier component of $\rho(x)$ is

$$2/d \int_{-\infty}^{\infty} \rho_a(x) \cos(2\pi m x/d) dx.$$

10. Without resorting to formal calculation, show that using a single frequency, a signal synthesised by combining all its harmonic frequencies with equal weight and phase will consist of a train of very sharp pulses separated by the period of the fundamental frequency.

11. Using a calculator or computer, complete the solution of the Square wave voltage problem of this Chapter and graph the resulting current $I(t)$.

25
The diffusion equation

In this chapter we shall tackle our first *partial* differential equation. This kind of equation involves partial derivatives (chapter 16) of a function with respect to different variables. In some respects, such as the consequences of linearity for the superposition of solutions, a partial differential equation is not very different from an ordinary differential equation (the kind which we have met thus far).

We shall see how a particular partial differential equation arises wherever a physical process is described as diffusion. Examples are heat transfer, the spreading of impurities in solids and the motion of neutrons in a reactor.

Heat is the most familiar example in practice, but one of the most subtle when it comes to fundamental understanding of the physics involved. We shall leave the discussion of such vexing questions as 'what is heat?' to physics courses and quote from them the following experimental result.

When a solid is set up as in fig. 25.1 so that temperature varies linearly in the x direction only, heat is conducted through it according to

$$\textit{Rate of transfer of heat} = -(\textit{cross-sectional area})$$
$$\times (\textit{thermal conductivity}) \times (\textit{temperature gradient}). \qquad (25.1)$$

The minus sign is convenient, making the thermal conductivity K positive, since heat flows from hotter to colder, i.e. in the opposite sense to the positive temperature gradient.

Suppose that the temperature T varies only in the x direction but not just linearly. We can use (25.1) to relate *local* heat transfer to *local* temperature gradient in order to describe this more general situation. To do so, consider a thin slab, of thickness Δx. Because the temperature gradient is different on its two faces, there is a net heat transfer into or out of this slab. The net amount of heat transferred to it after time t (starting at some arbitrary time) may be written $U A \Delta x$. (U is the energy per unit volume gained by the slab. It is objectionable, in physical terms, to call this the 'heat contained' in the slab). Application of the local form of (25.1),

$$\text{Rate of heat transfer} = - AK\frac{\partial T}{\partial x} \tag{25.2}$$

to the two sides of the thin slab gives us

$$A\,\Delta x\frac{\partial U}{\partial t} = - AK\left[\frac{\partial T}{\partial x}\bigg|_{x+\Delta x} - \frac{\partial T}{\partial x}\bigg|_{x}\right]. \tag{25.3}$$

Here the vertical lines and subscripts mean 'evaluated at'. Note that *partial* derivatives are involved because both U and T are, in principal, functions of *two* variables, position x and time t.

Our now familiar linear approximation $(f(x + \Delta x) - f(x) \approx \Delta x f'(x))$ can be used to rewrite this as

$$\frac{\partial U}{\partial t} = K\frac{\partial^2 T}{\partial x^2} \tag{25.4}$$

This is an equation for two different functions $U(x, t)$ and $T(x, t)$ but they

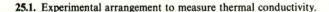

25.1. Experimental arrangement to measure thermal conductivity.

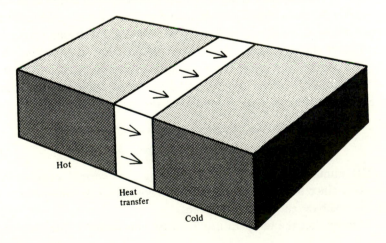

Hot

Heat
transfer

Cold

are related by another experimentally established relation (also used in chapter 18)

$$U = \rho C T + \text{constant}, \tag{25.5}$$

where ρ is the density and C the specific heat of the material. This simple linear relationship works well provided the range of temperature is not too large.

We can use this to eliminate U or T from (25.4). The equation for T is

$$\frac{\partial T}{\partial t} = \frac{K}{\rho C} \frac{\partial^2 T}{\partial x^2} \tag{25.6}$$

This is the partial differential equation which we sought, whose solution will tell us how temperature varies with time and position in our specimen. It is known as the *Diffusion Equation*. In this case, the constant $K/\rho C$ is called the thermal diffusivity. In other contexts, the equation will have some other physical constant playing the same mathematical role. Let us call it D, the diffusion constant, in general.

The manner in which we have derived the equation is typical of many such derivations in mathematical physics. One considers a small local element in space – a slice, or a small cube, or whatever – and determines the net rate of flow of some physical quantity such as mass, heat (energy), or momentum into that element.

Without (for the moment) trying to solve the equation, let us picture its solution in a simple case. This process should precede any detailed attack on a problem in physical mathematics, just as it should conclude with the question – is my result reasonable?

Suppose a thick slab is rapidly heated at one face, so that its temperature there is raised while it is unchanged in the interior. If no more heat is supplied and the surface is kept insulated, heat will be transferred to the interior progressively, as in our sketch, fig. 25.2. Note that heat is conserved in this case – it moves around but does not appear or disappear. This is justified (and generalised to cases where it is not true) by thermodynamics.

The word *diffusion* is also applied to the motion of a particle which moves randomly, as in the Brownian motion of a particle in a fluid. An impurity atom in a solid can diffuse at high temperatures, and the semiconductor industry uses large furnaces to encourage atoms to do so as an essential step in the manufacture of integrated circuits. But the electronics engineer is hardly concerned with the history of an individual atom. He introduces a concentration of impurities at the surface, and places the material in a furnace to 'drive in' the impurities. His concern is with the final concentration profile extending into the material. It turns out that this obeys the same equation (25.6) which we have been studying, and the process is directly analogous to the heat conduction problem envisaged above.

We shall not try to justify in full detail this connection between randomly moving particles and the diffusion equation, but the following rough arguments may help to justify it.

We divide our thick slab up into a finite number of thin ones, put a large number of particles (say, 1024) in the first slice and allow them to jump backwards and forwards at regular intervals. In fact, we shall make them jump backwards and forwards in equal numbers. Were we to go to sufficiently large numbers, this would amount to the same thing as equal probabilities for the two kinds of jump. In table 25.1, successive rows show how the population develops with time (the time unit is the interval between jumps). Initially ($t = 0$) the entire population of 1024 is accommodated near the face in cell 1 (the only significance of 1024 being that it can be repeatedly divided by 2). At $t = 1$ half of them jump to cell 2. They are unable to

25.2. Expected variation of temperature distribution with time.

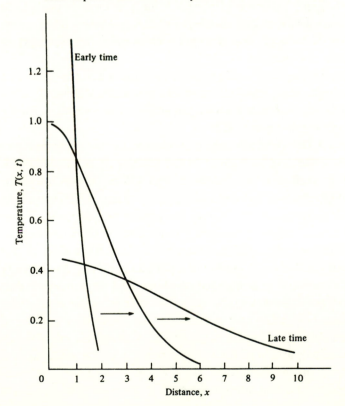

move to the left because we assume that the surface is well insulated. At $t = 2$ the population again changes with 256 returning to cell 1 and 256 progressing to cell 2. At $t = 3$, 128 jump forward to cell 3 while 128 jump backwards to cell 1. The pattern of splitting and combining is continued, with the result shown in table 25.1. Figure 25.3 shows the same results for $t = 5, 10$.

Several features are apparent. Since no particles are lost, the total

Table 25.1   *Occupation of cells in diffusion model, c(n).*

t	Cell number $c(n)$										
	1	2	3	4	5	6	7	8	9	10	11
0	1024										
1	512	512									
2	512	256	256								
3	384	384	128	128							
4	384	256	256	64	64						
5	320	320	160	160	32	32					
6	320	240	240	96	96	16	16				
7	280	280	168	168	56	56	8	8			
8	280	224	224	112	112	32	32	4	4		
9	252	252	168	168	72	72	18	18	2	2	
10	252	210	210	120	120	45	45	10	10	1	1

25.3. Graphical representation of some of the data of table 25.1.

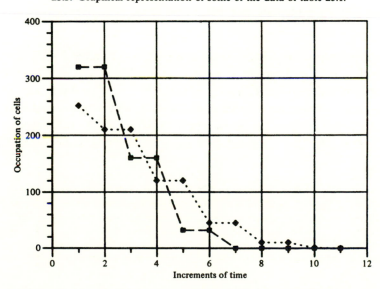

population, i.e. the sum of the row entries, must remain the same. In language appropriate to heat transfer, the hot face cools but deeper in there is an initial heating followed by a cooling. Incidentally, you may notice a relation between row 10 and our earlier table 2.1, which is worth thinking about. It makes one suspicious that, ultimately, our density profile must be a Gaussian function!

The mean square displacement of particles may be calculated from the occupation numbers $c(n)$ of table 25.1 as

$$d^2 = \sum_{n=1}^{\infty} (n-1)^2 c(n)/1024. \tag{25.7}$$

Here we have taken the width of each cell (or slice) as unity for convenience. The results are shown in table 25.2. A graph of these results will show clearly that the mean square displacement is proportional to the elapsed time. This is a general feature of diffusion processes. Since d is proportional to the square root of time, the spreading of the particles slows down as time proceeds.

But how is this calculation connected with our earlier discussion? Remember that we started out with continuous variables x and t, which require a population density function $N(x, t)$; its meaning is that $N(x, t)\Delta x$ is the population occupying a short section Δx. Then our rule becomes

$$\tfrac{1}{2}N(x - \Delta x, t) + \tfrac{1}{2}N(x + \Delta x, t) = N(x, t + \Delta t). \tag{25.8}$$

Subtracting $N(x, t)$ from both sides, it is supposed that the difference is very small compared to $N(x, t)$ itself, and that $N(x, t + \Delta t) - N(x, t) = (\partial N/\partial t)\Delta t$. The left-hand side can then be written

$$\tfrac{1}{2}[N(x - \Delta x, t) - N(x, t)] - \tfrac{1}{2}[N(x, t) - N(x - \Delta x, t)].$$

Each of these expressions is Δx times the partial derivative $\partial N/\partial x$ but evaluated first at x, and second at $x - \Delta x$. Their difference is $(\Delta x)^2$ times the *second* partial derivative, giving $\tfrac{1}{2}(\Delta x)^2 \partial^2 N/\partial x^2$. The final step is to assume that, with a large number of particles, Δx and Δt can be reduced indefinitely (which makes the above approximations exact) but keeping the ratio $D = \tfrac{1}{2}(\Delta x)^2/\Delta t$ constant. We then recover the same type of

Table 25.2

Time t	1	2	3	4	5	6	7	8	9	10
Mean square displacement d^2	0.50	1.25	2.00	2.81	3.63	4.47	5.31	6.18	7.04	7.91

equation as before

$$D\frac{\partial^2 N}{\partial x^2} = \frac{\partial N}{\partial t}. \tag{25.9}$$

Thus the random diffusing process which we envisage does indeed lead to the diffusion equation. This is the way that a physicist would think of it, being conscious of the fact that some random microscopic process always underlies the physics of diffusion. Heat conduction in particular can be thought of as the random diffusion of *energy*, exchanged in collisions between the agitated atoms of a solid. A mathematician might reverse the logic, regarding our simple arithmetic above as a step-by-step *approximation* to the solution of (25.9).

The relation (25.1), can also be recovered from our model with the limits taken above, in the form

$$\begin{matrix}\text{Rate of transfer at } x \\ \text{(in direction of positive } x)\end{matrix} = -D\frac{\partial N}{\partial x}. \tag{25.10}$$

In particle diffusion, it goes by the name of Fick's Law. This must be invoked to express appropriate boundary conditions when we seek an analytical solution, as we shall now do. For the problem at hand, there is no flow of particles across the end face. The boundary condition at the face $x = 0$ is

$$\left.\frac{\partial N}{\partial x}\right|_{x=0} = 0 \tag{25.11}$$

since there is no flow of N across it. For a finite slab a similar boundary condition would apply at the other end. Without such a boundary it is usually supposed that both N and $\partial N/\partial x$ tend to zero at large distances.

Conservation of particles (or no heat loss) may be demonstrated with the help of these conditions. The total number of particles (or amount of heat) is given by the integral $\int_0^\infty N\,dx$. Then from (25.9)

$$\frac{\partial}{\partial t}\int_0^\infty N\,dx = D\int_0^\infty \frac{\partial^2 N}{\partial x^2}\,dx = D\left[\left.\frac{\partial N}{\partial x}\right|_\infty - \left.\frac{\partial N}{\partial x}\right|_0\right] = 0.$$

It follows that $\int_0^\infty N\,dx$ is independent of time, and for normalised N it is unity.

The rate of diffusion (or conduction) is similarly obtained. As discussed above it is effectively the problem of the random walk which deals with the statistical progress of a particle equally likely to move forwards or backwards. In terms of the continuous normalised density $N(x, t)$, the mean value of the squared displacement x^2 is

$$\langle x^2 \rangle = \int_0^\infty N(x, t)x^2 \, dx. \tag{25.12}$$

Using (25.9) as before, and, integrating twice by parts,

$$\frac{\partial}{\partial t} \langle x^2 \rangle = D \int_0^\infty x^2 \frac{\partial^2 N}{\partial x^2} \, dx$$

$$= \left[x^2 \frac{\partial N}{\partial x} \right]_0^\infty - [2xN]_0^\infty + 2 \int_0^\infty N \, dx.$$

For the present problem, the first two right-hand terms vanish so that integrating with respect to time and using $\langle x^2 \rangle = 0$ at $t = 0$,

$$\langle x^2 \rangle = 2Dt. \tag{25.13}$$

Equation (25.13) was verified earlier for table 25.2.

The special solution of the diffusion equation corresponding to table 25.1 is denoted by $G(x, t)$ and takes the form

$$G(x, t) = (\pi Dt)^{-\frac{1}{2}} \exp(-x^2/4Dt). \tag{25.14}$$

It is worthwhile checking (*i*) that $G(x, t)$ is a solution – just substitute into (25.9); (*ii*) that it satisfies boundary condition (25.11); (*iii*) that it is normalised, i.e. that $\int_0^\infty G(x, t) \, dx = 1$ independently of time t; (*iv*) that $\langle x^2 \rangle = \int_0^\infty x^2 G(x, t) \, dt = 2Dt$. It is also worthwhile comparing the x and t dependence of $G(x, t)$ with the entries in table 25.1. The most remarkable behaviour of $G(x, t)$ occurs for small values of t. So long as t is not actually zero, at the boundary the exponential factor is identically unity so that $G(0, t) = (\pi t)^{-\frac{1}{2}}$. As t is reduced, $G(0, t)$ increases without limit. But just below the surface (finite x) the exponential factor takes over. The exponent $(-x^2/4Dt)$ becomes numerically large as t is reduced so that $G(x, t)$ is effectively extinguished except for a narrow region of width $(4Dt)^{\frac{1}{2}}$. Thus the product of height with width (and also the area) remains independent of time, as required by normalisation.

Thus as $t \to 0$, the x-dependence of $G(x, t)$ develops a singular behaviour – indefinitely large near $x = 0$, practically zero elsewhere and with an area equal to unity. This extraordinary function (it is not a function at all in a strict mathematical sense) finds considerable application in physics and engineering. It is termed the delta function, denoted by $\delta(x)$. Thus, in a shorthand notation

$$\begin{aligned} \delta(x) &= \infty, \quad x = 0 \\ &= 0, \quad\;\; x \neq 0 \end{aligned} \tag{25.15}$$

and

$$\int_a^b \delta(x)\,dx = 1, \quad a < 0 < b. \tag{25.16}$$

The integration range does not matter so long as it includes $x = 0$.

One is tempted to define the delta function directly as the limit $(z \to 0)$ of the function

$$\Delta_z(x) = (\pi z)^{-\frac{1}{2}} \exp(-x^2/4z) \tag{25.17}$$

which is illustrated in fig. 25.4. However this is not satisfactory because of the singular behaviour in that limit. The limit can only be taken *after* we perform some integral involving the delta function, i.e.

$$\int_a^b \delta(x) f(x)\,dx = \lim_{z \to 0} \int_{x_1}^{x_2} \Delta_z(x) f(x)\,dx.$$

In this way sanity and mathematical respectability are restored.

It must not be thought that $G(x, t)$ is the only solution of the diffusion equation. Heat conduction could start from any source, e.g. the centre of a laser beam, presenting a more awkward cylindrical symmetry. Moreover, mathematically speaking, 'initial' could refer to any instant – the value $t = 0$ has no special significance for the diffusion equation itself.

25.4. The Incredible Shrinking Function
$(\pi z)^{-1/2} \exp(-x^2/4z)$
for $z = 1, 10, 100, 1000$.

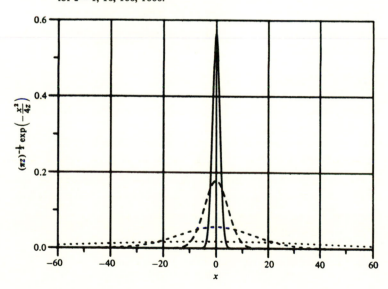

Summary

Diffusion of particles or heat in one dimension is described by the partial differential equation,

$$\frac{\partial N}{\partial t} = D \frac{\partial^2 N}{\partial x^2}.$$

The boundary condition at a plane surface is

$$\left(\frac{\partial N}{\partial x} \right)_{\text{surface}} = 0.$$

A special solution referring to an initial concentration at $x = 0$ is

$$G(x, t) = (\pi D t)^{-\frac{1}{2}} \exp(-x^2/4Dt).$$

In the limit $t \to 0$, $G(x, t)$ reduces to the delta function, $\delta(x)$, in the sense described in the text.

EXERCISES

1. Plot $G(x, t)$ (25.14), taking $D = 0.5$, as a function of x for $t = 8.0$ and compare with the corresponding entries in table 25.1.

2. Plot $G(x, t)$ as a function of t for $x = 3.0$ and compare with the normalised values obtained from table 25.1.

3. Construct a diffusion table describing 2048 particles initially concentrated at the centre of an unlimited medium, i.e. able to diffuse both left and right. Discuss the corresponding solution of the diffusion equation.

4. Verify that $G(x, t)$ (25.14) is a solution of the diffusion equation (25.9). By differentiation with respect to t show that

$$(\pi D t)^{-\frac{1}{2}} \int_0^\infty x^2 \exp(-x^2/4Dt) dx = 2Dt.$$

5. Verify by direct substitution that a solution of the diffusion equation is
$N(x, t) = A[1 - \exp(-D\alpha^2 t) \cos \alpha x]$
where A and α are constants. Sketch $N(x, t)$ (*i*) as a function of x for various t; (*ii*) as a function of t for various x.

6. If the normalisation condition $\int_{-L}^{L} N(x, t) dx = 1$ is imposed on the function in exercise 5, what is the smallest possible value of L? Evaluate A.

7. Show that $\int_{-\infty}^{\infty} G(x', t_1) G(x - x', t_2) dx' = G(x, t_1 + t_2)$.
   [Hint: use the identity

$$\frac{x'^2}{t_1} + \frac{(x - x')^2}{t_2} = \left(\frac{1}{t_1} + \frac{1}{t_2} \right) \left(x' - \frac{t_1 x}{t_1 + t_2} \right)^2 + \frac{x^2}{t_1 + t_2}$$

8. Show that a general solution of the diffusion equation is

$$N(x, t) = \int_{-\infty}^{\infty} N(x', 0)G(x - x', t)\mathrm{d}x'.$$

9. The linearity of the diffusion equation enables us to combine solutions additively, in such cases as the following. A certain concentration of impurity is introduced at the surface of a semiconductor wafer and diffused into it for four days at a high temperature. The whole operation is then repeated exactly on the same wafer, with a diffusion time of one day. Make a careful *sketch* of the shape of the resulting concentration profile.

10. Show (not necessarily with full mathematical rigour) that the δ function has the property

$$\int_{x_1}^{x_2} f(x)\delta(x - x_0)\mathrm{d}x = f(x_0), \quad \text{provided } x_1 < x_0 < x_2.$$

What is the value of the integral if x_0 is outside the range x_1 to x_2? Express these results in words and memorise them, as a basic working definition of the delta function to be used whenever it crops up inside an integral (which is, in terms of its definition, where it belongs!).

26
Waves

Confronted with an expanse of still water in a pond, who can resist the temptation to do the obvious experiment, to fling a stone into its centre? The disturbance thus created travels outwards, is reflected at the pond's edge and eventually disappears. Equilibrium is restored.

Such a disturbance is called wave motion. It is similar to the oscillatory motion that we have studied in chapters 21 and 23, except that it describes the behaviour of an extended medium (the surface of the pond) whose equilibrium is disturbed. The appropriate mathematical description must involve a function of several variables, since the height of the pond's surface is a function of position (x, y) as well as time (t). Already we can expect that a partial differential equation may be the key to understanding it.

Physicists are always striving for unified descriptions of the physical world. In the nineteenth century a variety of apparently different phenomena came to be understood in terms of waves – elastic waves in solids, sound waves in air, and, most importantly, electromagnetic waves. These include light, x-rays, radio waves, infrared radiation and so on, and constitute in themselves another important 'unification' of physics. All of these types of waves and many others have a common mathematical description, at least up to a point.

One of the great surprises of modern physics in this century was the realisation that matter itself also has wave-like properties, described by quantum mechanics.

Energy is required to generate waves, as when we hurl the stone into the pond. The dissipation of this energy (usually as heat) is what causes the diminishing or attenuation of waves, just as the oscillations described in chapter 23 were attenuated. If this effect is not too great, waves can be used to transmit energy, or information, from a source which generates the waves to a receiver which absorbs them.

Waves on the open sea are in confused, ever-changing motion – the product of myriad distant storms and breezes. Sounds at a cocktail party or ordinary white light are similarly complicated. How could we ever come to grips mathematically with such things? The answer is similar to (and related to) that given for the representation of arbitrary functions as Fourier series, chapter 24. We can consider any wave motion to be a sum of elementary *sinusoidal* wave motions which are readily understood.

The traditional simple one-dimensional example is the stretched string (fig. 26.1), because position is specified by a single variable x. We assume that the displacement of the string, which is the quantity which varies in the wave motion, is confined to one plane, the plane of the paper in our sketch. We shall write this as $\psi(x, t)$ (Greek *psi*).

When the string is disturbed its tension pulls it back towards its equilibrium and is responsible for the wave motion. The laws of mechanics (essentially just *force = mass × acceleration* applied locally on the string)

26.1. Vibration of stretched string.

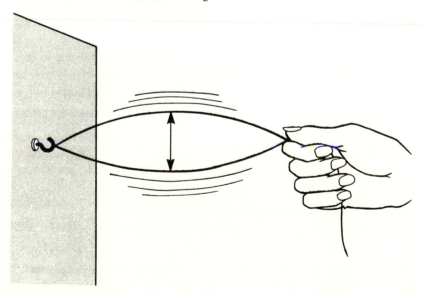

dictate that, provided the disturbance is small and frictional or viscous effects (and hence attenuation) are neglected,

$$\frac{\partial^2 \psi}{\partial x^2} = v^{-2}\frac{\partial^2 \psi}{\partial t^2}. \tag{26.1}$$

This is our expected partial differential equation, which was derived by Jean Le Rond d'Alembert in 1747. A derivation, based on examining a small element of the string, is usually provided in introductory physics texts. The constant v is defined by

$$v = (T/\mu)^{\frac{1}{2}} \tag{26.2}$$

where T is the tension force and μ is the mass per unit length of the string.

Equation (26.1) is the *Wave Equation*. In appearance it is rather similar to the Diffusion Equation (25.6) but the underlying physics is quite different – oscillation rather than diffusion. These are two complementary aspects of the real world which must ultimately be combined in physical theories.

Our strategy here is to find the simplest solutions of (26.1) and then build up our theory in terms of these elementary waves.

The simplest type of solution is just the *sinusoidal wave*

$$\psi = A \sin[k(x \pm vt) + \phi] \tag{26.3}$$

where A, k and ϕ are arbitrary constants. This may be verified by substitution. Note its significance (fig. 26.2). When frozen in time, the wave is just a sine function of x, but the function moves along in the x direction with velocity v as time passes. It moves to the right if we choose the negative sign, to the left if we choose the positive sign in (26.3). It is sometimes called a *travelling* wave for this reason.

This functional form is simple, but it can be written in various ways, so that there is a long list of definitions to be learned at this point. These are:

A—amplitude

k—wavenumber (or wave vector, because it becomes a vector later)

v—velocity (or phase velocity)

ϕ—phase constant – often omitted for purposes of discussion

$\omega = kv$ – circular frequency (units: radians per second)

$\lambda = 2\pi/k$ – wavelength

$\nu = \omega/2\pi$ – frequency (units: $\sec^{-1}$, or Hz)

(N.B. $v = \nu\lambda$)

Note that by changing the phase constant ϕ, (26.3) can equally well be expressed in terms of the cosine function. For some purposes, it is more convenient to use the complex solution

$$\psi = A \exp[ik(x \pm vt) + i\phi]$$

in the spirit of chapter 22, and eventually take its real part.

With a real stretched string, it is not easy to generate anything like (26.3) because it is usually clamped at both ends. This boundary condition is expressed mathematically as

$$\psi(o, t) = \psi(L, t) = 0, \quad \text{for all } t, \tag{26.4}$$

and it cannot be satisfied by (26.3), unless we just put $A = 0$.

However, the combination of just two sinusoidal travelling waves is sufficient to provide a possible wave motion of this system. These two waves are defined by the two choices of sign in (26.3), and are summed to give

$$\psi(x, t) = A[\sin k(x - vt) + \sin k(x + vt)]. \tag{26.5}$$

26.2. Travelling wave at various times.

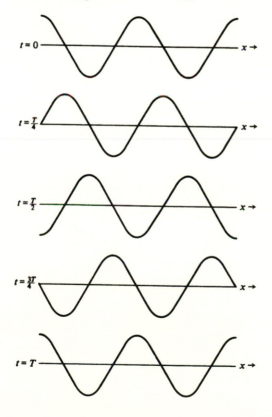

It is obvious that this is still a solution of (26.1). In fact, any number of sinusoidal waves (with any amplitude, wave vector or phase constant) can be combined. This is the *principle of superposition*, which follows, just as it did for ordinary differential equations, from the fact that the Wave Equation (26.1) is *linear* in ψ and its derivatives. Equation (26.5) may now be written as

$$\psi(x, t) = 2A \sin kx \cos \omega t, \tag{26.6}$$

and we see that it will indeed satisfy the required boundary conditions, provided

$$k = n\pi/L, n \text{ an integer.} \tag{26.7}$$

Such a solution is a *standing wave*—it oscillates up and down as in fig. 26.3. Remember that v is fixed, so each choice of n (and hence k, or wavelength λ) defines a standing wave with a different circular frequency ω (or frequency v), according to $\omega = kv$ or $v = v\lambda$.

26.3. Standing wave at various times.

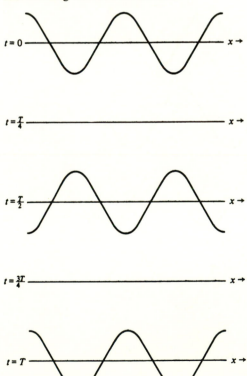

As a second example of the principle of superposition, consider the combination of two sinusoidal waves of slightly different frequency, travelling in the same direction. This may be written

$$\psi(x,t) = A \sin[(k - \Delta k)(x - vt)] + A \sin[(k + \Delta k)(x - vt)]$$
$$= 2A \sin k(x - vt)\cos\Delta k(x - vt). \tag{26.8}$$

The first factor, in isolation, is just a single sinusoidal wave, but it is multiplied by a second factor which takes the form of another sinusoidal wave with a very long wavelength. The result is as in fig. 26.4 – a wave whose amplitude varies to give 'beats'. Remember that this whole pattern will move steadily to the right, so if we were concerned with sound waves, an observer would hear a regular sequence of maxima in the intensity of sound.

There is clearly no end to this process of building up more and more complicated waves by superposition. Conversely, observed wave patterns can be broken down into sinusoidal components by the process of Fourier analysis, as described later.

There is one other kind of elementary wave that is worth mentioning. Since any function $f(x - vt)$ is easily seen to be a solution of (26.1), a simple *pulse* as in fig. 26.5, will propagate unchanged. In modern digital electronics, such pulses are the basis of communication.

We must be careful in indicating the degree of generality of what we have

26.4. An example of a beat pattern:
$\sin x + \sin 1.1x = 2 \sin 1.05x \cos 0.05 x$

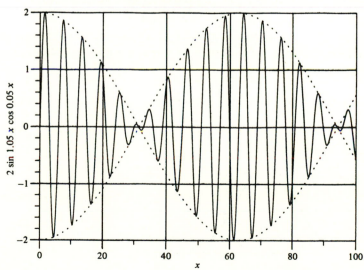

said about waves. The word 'wave' can be used whenever solutions of the form (26.3) (or combinations thereof) occur in physical systems, but they are not always solutions of an equation quite as simple as (26.1). The main effect of this is to make v a function of k, so that it is no longer a constant in the relation $\omega = kv$ or $v = v\lambda$. In such a case, waves of different wavelength or frequency travel at different speeds. This is called *dispersion*. For example, electromagnetic waves in a vacuum have no dispersion but, when propagating through matter, waves of different frequencies are slowed down to different extents. Ultimately, this is the cause of the well-known property of a prism, which 'disperses' white light into its constituent colours (wavelengths), emerging at different angles. Ordinary water waves, which are such a favourite illustration, exhibit strong dispersion. Note that the general form $f(x - vt)$ is no longer a solution in this case. In particular, a pulse will eventually break up because its constituent sine waves travel with different velocities and after a while they will fail to combine in the same way to make up the pulse.

Dispersion does not prevent us using the principle of superposition (provided the appropriate equation, replacing (26.1), is still linear). We must simply remember not to assume that v is a constant.

Further consideration of waves would require the full three-dimensional version of (26.1). Some of the exercises point in that direction but we shall confine ourselves to one dimension here and complete the illustration of wave properties with an example which may appeal to guitar players. It combines the ideas of this chapter and chapter 24 on Fourier series. We shall consider a string which is plucked at its midpoint and released, at $t = 0$. What is the displacement $\psi(x, t)$ of the string at subsequent times?

The simple standing wave solutions (26.6), already derived, provide the means to the construction of a solution. We write a suitable combination of these standing waves as

$$\psi(x, t) = \sum_{n=1}^{\infty} C_n \sin(n\pi x/L) \cos(n\pi vt/L). \qquad (26.9)$$

26.5. Travelling pulse.

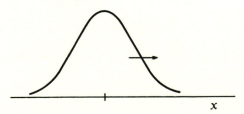

x

The question is, what are the coefficients C_n? They are determined by the initial conditions, that $\psi(x, 0)$ take the form

$$\psi(x, 0) = 2ax/L, \quad 0 \leqslant x \leqslant L/2$$
$$= 2a(L - x)/L, \quad L/2 \leqslant x \leqslant L \tag{26.10}$$

Putting $t = 0$ in (26.10) we have

$$\psi(x, 0) = \sum_{n=1}^{\infty} C_n \sin(n\pi x/L) \tag{26.11}$$

so the coefficients C_n are just the Fourier coefficients of the function (26.10) (or rather the function which repeats its form periodically over intervals of length $2L$ – but only the portion $0 \leqslant x \leqslant L$ is relevant). We have already found these coefficients in exercise 7 of chapter 24.

This strategy is a common one in the solution of linear partial differential equations, as an alternative to the kind of numerical method described in the last chapter. The form of solution (26.9) may look complicated, but it is really quite manageable for computational purposes.

Summary

Sinusoidal travelling waves have the form

$$\psi = A \sin[k(x \pm vt) + \phi]$$

where A = amplitude, k = wavenumber, v = velocity, ϕ = phase constant, and $\omega = kv$. They are generally solutions of a linear partial differential equation, which allows them to be combined (superposition principle). In particular the combination of $+$ and $-$ in the above gives standing waves. Waves which obey the simple equation

$$\frac{\partial^2 \psi}{\partial x^2} = v^{-2} \frac{\partial^2 \psi}{\partial t^2}$$

have constant v, but many important physical waves have $v = v(k)$ (dispersion). Superposition + Fourier analysis $\rightarrow$ solution as a series of sinusoidal waves.

EXERCISES

1. Discuss the form of the wave which is formed when one of the two travelling waves in (26.5) is given an amplitude twice as large as the other. This is what might happen when a wave is partially reflected at some point and the incident and reflected waves combine.

2. What are the mean and mean square values of ψ, as given by (26.3)(i) taken

with respect to x for fixed t (*ii*) vice-versa? In physics, the mean square value generally gives the *intensity* of the wave.

3. Compute the form of the solution (26.10) at the end of one period of the 'fundamental mode' which is the $n = 0$ sinusoidal standing wave. Sketch it and discuss its shape in terms of the contribution of the first few terms in the solution.

4. The most common types of boundary condition are:

   (a) $\psi = 0$
   (b) $\psi' = 0$ $\Big\}$ some combination of these at $x = 0$ and $x = L$

   or

   (c) $\psi(0) = \psi(L)$
   $\psi'(0) = \psi'(L)$ $\Big\}$ periodic boundary conditions.

   Find a formula for the frequency of the fundamental mode for each combination of (a) and (b) and for (c). Note that, in the case of periodic boundary conditions, it is as if the physical system consisted of a ring of perimeter L.

5. A long train consists of railway carriages all of mass M connected by springs (Hooke's constant λ). Set up the equations which govern the coupled motions of the carriages. Relate these to the wave equation, (something similar was done for diffusion in the last chapter).

6. Two equal 'pulses' (see text) with opposite sign propagate in opposite directions. Describe what happens when they pass one another, with sketches of ψ and its time derivative.

7. Marconi sent the first trans-Atlantic radio signal, to the great consternation of physicists who claimed that it was not possible. Actually he is thought to have been lucky – the signal was carried by the 'first harmonic'. Explain what this means in terms of this chapter and chapter 24.

8. A step-by-step numerical method is applied to the solution of (26.1), representing interval $(0, L)$ by N points, and using time increments Δt. If the method is intended to describe motion which consists mainly of the fundamental and first ten harmonics (for boundary conditions $\psi(0) = \psi(L) = 0$), what are the requirements regarding N and Δt for reasonable accuracy?

9. (Modulation of a wave)
   The wave represented by
   $\psi = A \cos(kx - \omega t + \phi)$
   is said to be modulated in amplitude or phase if A or ϕ is given a time-dependence (usually such that its variation is on a much longer time scale than the period of the original wave).

Suppose that in each case (considered separately) the time-dependent quantity is taken to be $a + b \cos \Delta\omega t$. Show that the modulated wave can be decomposed into a combination of sinusoidal waves as follows:

(a) for amplitude modulation, waves of frequency $\omega, \omega \pm \Delta\omega$;

(b) for phase modulation, waves of frequency $\omega, \omega \pm \Delta\omega$ (neglecting terms of order b^2).

What are the amplitudes of the various components?

10. (Frequency modulation) The wave represented by

$\psi = A \cos(kx - \omega t + \phi)$

is said to be modulated in frequency if ω is made time dependent. Show that such a wave modulated according to $\omega = \omega_0 + b \cos \Delta\omega t$ can be decomposed into sinusoidal waves of frequency $\omega_0, \omega_0 \pm \Delta\omega \pm b$ (neglecting terms of order b^2).

What are the amplitudes of the various components?

27

The rate of change of a vector

Nineteenth-century texts in applied mathematics have a 'busy' appearance. An equation for each dimension seems to be the rule, demanding a Victorian abundance of energy and leisure. Modern texts use the labour-saving device of the vector in one form or another. Comprehensive statements are abbreviated to just a few significant symbols; these, however, must be treated with considerable respect.

An example will help to make this clear. In chapter 20, we examined the oscillations of a particle placed in a bowl. However, as well as motion in a vertical plane, say that containing the x-axis, more complicated motion is possible. Two equations are required to express Newton's laws for the variation of the x and y coordinates, which might be written

$$\frac{dv_x}{dt} = -\omega^2 x$$

$$\frac{dv_y}{dt} = -\omega^2 y. \tag{27.1}$$

Here y is the horizontal coordinate in a second vertical plane $(x = 0)$ and v_x, v_y are velocity components in the x and y directions.

By putting $dv_x/dt = d^2x/dt^2$ etc. two independent linear second-order equations are obtained for x and for y. Their general solution, requiring $2 \times 2 = 4$ constants of integration, describe the possible motion of the particle. For example $x = R \cos \omega t$, $y = R \sin \omega t$ specify a circular path at a distance R from the centre (because $x + y^2 = R^2$ independently of t). More complicated motions can easily be imagined.

The pair of equations (27.1) could be written with fewer symbols. As described in chapters 3 and 4, the pair of coordinates (x, y) form the components of the position vector **r**, namely $\mathbf{r} = x\mathbf{i} + y\mathbf{j}$, where **i**, **j** are the unit vectors along the fixed x, y axes. In our example a vector **r** (position) is a function of a scalar t (time). Similarly, (v_x, v_y) are the components of the velocity vector **v**, that is, $\mathbf{v} = v_x\mathbf{i} + v_y\mathbf{j}$. Thus, multiplying the first and second parts of (27.1) by **i**, **j** respectively, and adding, both equations are summarised by the single vector equation,

$$\frac{d\mathbf{v}}{dt} = -\omega^2 \mathbf{r} \tag{27.2}$$

or

$$\frac{d^2\mathbf{r}}{dt^2} = -\omega^2 \mathbf{r}. \tag{27.3}$$

Note that the derivative of a vector has quite an elementary definition, namely (using three dimensions, for generality)

$$\frac{d}{dt}\mathbf{a} = \frac{da_x}{dt}\mathbf{i} + \frac{da_y}{dt}\mathbf{j} + \frac{da_z}{dt}\mathbf{k}. \tag{27.4}$$

Using this, the rules for differentiating a vector are easily established. An important one concerns the scalar product (chapter 5). Thus

$$\frac{d}{dt}(\mathbf{a}\cdot\mathbf{b}) = \dot{a}_x b_x + a_x \dot{b}_x + \dot{a}_y b_y + a_y \dot{b}_y + \dot{a}_z b_z + a_z \dot{b}_z$$

$$= \mathbf{a}\cdot\frac{d\mathbf{b}}{dt} + \frac{d\mathbf{a}}{dt}\cdot\mathbf{b}. \tag{27.5}$$

The resemblance of (27.5) to the usual rule for differentiating a product will be obvious.

As well as the algebraic definition (27.4) there is an equivalent geometric

27.1. Successive positions P and Q and corresponding vectors.

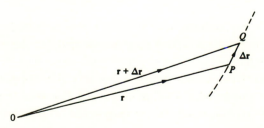

meaning. This is illustrated in fig. 27.1 for the rate of change of **r**, namely **v** = d**r**/d*t*. The vector **r** = *x***i** + *y***j** + *z***k** is drawn from some origin *O*. The choice of *O* may be important in practice but it does not affect the ensuing definition. An any instant *t* the position of the particle is designated by the vector **r**, represented by *OP*. Assuming a smooth dependence of **r** (i.e. of the coordinates *x*, *y*, *z*) upon *t* then at a later instant *t* + Δ*t*, the particle has moved to a new position, **r** + Δ**r**. As Δ*t* is reduced to very small values, the ratio Δ**r**/Δ*t* tends to some definite limiting (vector) value. This is the velocity **v**. Thus

$$v = \lim_{\Delta t \to 0} \frac{\Delta r}{\Delta t} = \frac{dr}{dt}. \tag{27.6}$$

Of course, using the components *x*, *y*, *z* of **r**, **v** can also be written in a form analogous to that of (27.4) as

$$v = \frac{dx}{dt}i + \frac{dy}{dt}j + \frac{dz}{dt}k. \tag{27.7}$$

Perhaps the most important feature of (27.6) is that Δ**r** is not in the same direction as **r** (in general), and therefore d**r**/d*t* is not necessarily parallel to **r**. In fact, it is clear from fig. 27.1 that it is in the direction of the *tangent* to the curve defined by **r**(*t*).

As an illustration, consider motion on the circle $x^2 + y^2 = R^2$. This could be expressed vectorially as $\mathbf{r} \cdot \mathbf{r} = R^2$. Using rule (27.5), and remembering that *R* does not depend on time,

$$r \cdot \frac{dr}{dt} = r \cdot v = 0.$$

According to chapter 5, this result implies that **v** is perpendicular to **r** for circular motion about the origin. This is just the familiar property of a circle, that any radius is perpendicular to the tangent as its end-point. There is a further consequence of this result, if the circular motion obeys (27.2). Taking the scalar product of both sides of (27.2) with **v** gives (for $\mathbf{r} \cdot \mathbf{r} = R^2$)

$$v \cdot \frac{dv}{dt} = - \omega^2 v \cdot r = 0$$

Thus, integrating, $\mathbf{v} \cdot \mathbf{v} = \text{constant}$, that is to say, the magnitude of the velocity is constant in time, as indeed may be verified for the explicit solution $\mathbf{r} = (R \cos \omega t, R \sin \omega t)$.

According to chapter 6 in a small rotation Δ*ϕ* of an object about an axis through the origin, any point with position vector **r** is displaced to **r** + Δ**r**, where Δ**r** = Δ*ϕ* × **r**. As explained, Δ*ϕ* has the magnitude Δ*ϕ* of the rotation, and the direction of the rotation axis. This expression can now be used directly to obtain the velocity of the point **r**. Thus dividing Δ**r** by the (small) time interval Δ*t* gives lim(Δ**r**/Δ*t*) = d**r**/d*t*, so that

$$\mathbf{v} = \frac{d\mathbf{r}}{dt} = \boldsymbol{\omega} \times \mathbf{r} \tag{27.8}$$

where $\boldsymbol{\omega} = \lim_{\Delta t \to 0} (\Delta \phi / \Delta t)$ is called the *angular velocity* vector.

Equation (27.8) is deceptively simple, the catch being that not only are **r**, **v** both time-dependent, but in most situations of physical interest so is **ω**!

Summary
The derivative of vector $\mathbf{a} = a_x \mathbf{i} + a_y \mathbf{j} + a_z \mathbf{k}$ is defined as

$$\frac{d\mathbf{a}}{dt} = \frac{da_x}{dt}\mathbf{i} + \frac{da_y}{dt}\mathbf{j} + \frac{da_z}{dt}\mathbf{k}.$$

Differentiation of scalar product:

$$\frac{d}{dt}(\mathbf{a} \cdot \mathbf{b}) = \mathbf{a} \cdot \frac{d\mathbf{b}}{dt} + \frac{d\mathbf{a}}{dt} \cdot \mathbf{b}.$$

Rotation for rigid body rotating with angular velocity **ω**, velocity **v** = d**r**/d*t* at any point *r* is given by

$$\mathbf{v} = \boldsymbol{\omega} \times \mathbf{r}$$

EXERCISES

1. Denoting the time derivative of the position vector **r** by d**r**/d*t* = **v**, evaluate the rate of change of (*i*) **c**·**r**, (*ii*) *c***r**, (*iii*) $|\mathbf{r}|^2$, (*iv*) $\mathbf{r}/|\mathbf{r}|^2$ (**c** is time-independent).

2. Derive a formula analogous to (27.5) for the derivative of **a** × **b**.

3. The force on a particle is denoted by **F** so that the equation of motion is *m* d**v**/d*t* = **F**. Assuming that **F** is parallel to **r**, show that (d/d*t*) (**r** × **v**) = 0 ('conservation of angular momentum').

4. Work out **v** = (d**r**/d*t*) and **a** = (d**v**/d*t*) for the case
   r = (*R* cos *ωt*)**i** + (*R* sin *ωt*)**j**
   and comment on the values of **r**·**r**, **r**·**v**, **r**·**a** and **v**·**a**.

5. A particle is located at the point $\mathbf{r}_0$ at time *t* = 0 and has constant velocity **v**. What is the equation for its path?

6. A wheel of radius *R* rolls steadily in a straight line along a flat surface. Write a single equation for the position of one point on the rim, as a function of time. Hence derive an equation for the velocity of the point.

7. If the motion described in exercise 4 is in the (*x*, *y*) plane, what is the direction and magnitude of the angular velocity vector for this motion?

8. The equation for the centre of mass of a body of total mass M may be expressed in a somewhat abbreviated notation as
$$M\mathbf{R} = \Sigma m\mathbf{r}.$$
Use this to show that if the axis of rotation passes through the centre of mass, the total momentum $\Sigma m\mathbf{v}$ associated with any rotational motion is zero.

9. Derive the following formula for the acceleration of a point fixed in a rotating body, using the formulae given in the text.
$$\mathbf{a} = \boldsymbol{\omega} \times (\boldsymbol{\omega} \times \mathbf{r}) + \dot{\boldsymbol{\omega}} \times \mathbf{r}$$

10. A particle describes a path $\mathbf{r}(t)$. This follows a curve which may also be expressed as $y(x)$. What is the relation between dy/dx and the derivatives of $\mathbf{r}$? Illustrate your answer with the example $\mathbf{r} = t\mathbf{i} + t^2\mathbf{j}$.

28

The scalar field and gradient operator

If the temperature of the air in a room varies from one point to another, it may be regarded as a function of position. Here we have a scalar (temperature) which is a function of a vector (position). It may be written as $\phi(\mathbf{r})$. In physics this kind of function is called a *scalar field*, and its most important application is to potentials, as we shall see in chapter 33.

Such a field is a little difficult to visualise in three dimensions. Most of the relevant mathematical apparatus can be better appreciated when $\mathbf{r}$ is two-dimensional. In physical geography, the mapping of terrain is just such a case – height z is a function of position (x, y) or $\mathbf{r}$. The use of contours – lines of equal height – is a familiar way of representing $z(\mathbf{r})$, as in fig. 28.1. The contours are defined by

$$z(\mathbf{r}) = \text{constant.} \tag{28.1}$$

and have already been the subject of some discussion in chapter 16.

Every functional relationship in applied mathematics invites the definition of a derivative. In the present case, we have already learned how to take two different derivatives of $z(\mathbf{r})$ – its *partial* derivatives with respect to x and y (chapter 16). But to do business with these separately would be to lose much of the efficacy of vector methods. Instead we usually represent them together by a single vector quantity, which tells us how z varies around (x, y). This is

$$\nabla z = (\partial z/\partial x, \partial z/\partial y) \tag{28.2}$$

which is called the *gradient* of z. It may be regarded as the result of operating upon z with the *gradient operator* $(\partial/\partial x, \partial/\partial y)$, which plays a role analogous to d/dx in ordinary derivatives.

The gradient is itself a function of **r**, but let us for the moment concentrate on some particular point and, remaining with our two-dimensional geographical example, examine its significance. Suppose that we measure height z on a path **r**(t) which passes through a given point – e.g. by use of a satellite which passes overhead (Fig. 28.1). What is the instantaneous rate of change of the measured height, i.e. dz/dt? The height z depends on **r**, which in turn depends on t – just the sort of 'chain' relationship considered in chapter 16, when we developed chain rules for partial derivatives.

In the present case, the required derivative may be expressed as

$$\frac{dz}{dt} = \frac{\partial z}{\partial x}\frac{dx}{dt} + \frac{\partial z}{\partial y}\frac{dy}{dt} \tag{28.3}$$

28.1. Contours of a function $z(x, y)$ and a path **r**(t) in the (x, y) plane.

which may be written very neatly as

$$\frac{dz}{dt} = \mathbf{v} \cdot \nabla z \tag{28.4}$$

where $\mathbf{v} = d\mathbf{r}/dt$. The rate of change of the measured height depends on the relative directions of $\mathbf{v}$ and ∇z and is a maximum when they are parallel. So ∇z 'points uphill' – it gives the magnitude and direction of the local *slope* of $z(\mathbf{r})$. Common sense, familiar to mountaineers, suggests that should be perpendicular to the contour which passes through the chosen point, as in fig. 28.2. If $\mathbf{r}(t)$ is chosen to follow the contour, the height does not vary, so

$$\mathbf{v} \cdot \nabla z = 0 \tag{28.5}$$

This indeed says that Δz is perpendicular to the tangent to the contour, which is what we mean when we say that it is perpendicular to the contour. (See also exercise 1).

All of this is mathematically similar, but harder to picture, in three dimensions. Returning to our example of the temperature in a room, the contours are now surfaces defined by constant temperature. The gradient $\nabla \phi$ is normal to these surfaces (i.e. perpendicular to any tangent to the surface through the chosen point). A bumble bee, executing a flight in a path $\mathbf{r}(t)$, would sense an instantaneous variation of temperature given by

$$\mathbf{v} \cdot \nabla \phi \tag{28.6}$$

Figure 28.3 shows an example of contour surfaces in three dimensions.

Points at which $\nabla \phi = 0$ are particularly interesting and are called *stationary points* (cf. chapter 12). These include local maxima and minima (fig. 28.4*a*, *b*) and saddle points (fig. 28.4*c*). Figure 28.5 shows a function with stationary points of various kinds. Much of modern engineering and economics is concerned with finding maxima and minima in some abstract space of variable design parameters. Sophisticated 'optimisation' programs have been developed which try to follow ϕ uphill or downhill to find these

28.2. The gradient is perpendicular to the direction of contour lines.

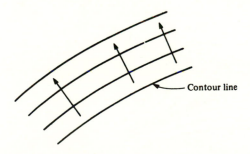

Contour line

28.3. Contour surfaces of constant density in a galaxy (J. Cantrella and A. Mellot, Lawrence Livermore Laboratory).

(*a*)   (*b*)

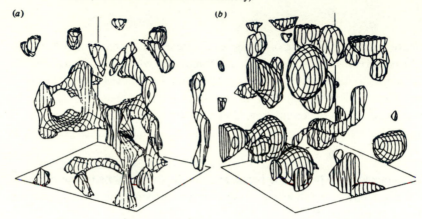

28.4. Form of contours around stationary points.

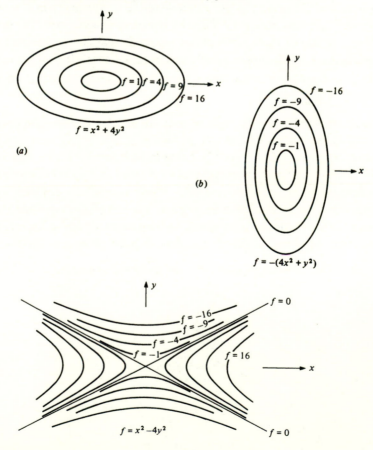

$f = x^2 + 4y^2$

(*a*)

(*b*)

$f = -16$

$f = -9$

$f = -4$

$f = -1$

$f = -(4x^2 + y^2)$

$f = 0$

$f = -16$

$f = -9$

$f = -4$

$f = -1$

$f = 16$

$f = x^2 - 4y^2$

$f = 0$

points, to optimise profit or efficiency. Of course, in this case the number of variables may well be much more than three! In physics, when we deal with potentials, we shall also attach special significance to stationary points, as points of equilibrium.

The rules for manipulation of $\nabla(\phi_1\phi_2)$ and $\nabla(\phi_1 + \phi_2)$ are fairly obvious if we recall what they represent – when in doubt, write out components. In particular

$$\nabla(c\phi) = c\nabla\phi \tag{28.7}$$

when c is a constant.

Earlier, we wrote ∇ as a vector operator and this idea is easily generalised to three dimensions, writing $\nabla = (\partial/\partial x, \partial/\partial y, \partial/\partial z)$. This suggests that it might be usefully combined with the $\cdot$ and $\times$ notations of chapters 5 and 6, to define new operators $\nabla\cdot$ and $\nabla\times$, and we shall do this in chapters 33 and 34 respectively.

Summary

A scalar function $\phi(\mathbf{r})$ of a vector may be called a scalar field. Contour lines (in two dimensions) or surfaces (in three dimensions) may be defined by setting the function equal to constant values. The local variation of ϕ is given by $\nabla\phi$ where ∇ is the gradient operator. $\nabla = (\partial/\partial x, \partial/\partial y, \partial/\partial z)$ in three dimensions.

$\nabla\phi$ is perpendicular to the lines or surfaces defined by constant ϕ.

$\nabla\phi = 0$ defines stationary points, including maxima, minima and saddle points.

28.5. Computer plot of function with stationary points.

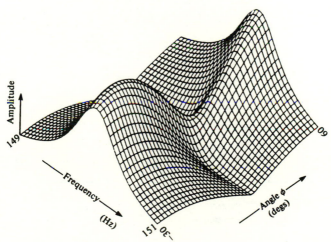

EXERCISES

1. Give an alternative proof of the statement that $\nabla\phi$ is perpendicular to the contour line through a given point, by taking the derivative of (28.1) with respect to x, under the condition that y is a function of x, given by (28.1) itself. This is a neater but logically intimidating proof! See chapter 16.

2. Give derivations of rules for the reduction of $\nabla(\phi_1\phi_2)$ and $\nabla(\phi_1 + \phi_2)$.

3. Sketch the contours of the scalar field
   $$\phi = \mathbf{a}\cdot\mathbf{r}$$
   where $\mathbf{a}$ is the constant vector $(1, 2)$. Show that $\nabla\phi = \mathbf{a}$.

4. *Sketch* the contours for the functions $x^2 - y^2, xy$, and $ax^2 + by^2$. Identify the stationary points in each case.

5. Experimental data in astronomy often consists of the intensity of radiation (e.g. infrared) from each of a grid of points in the sky (see fig. 28.6). Outline a suitable procedure for computing approximate intensity contours based on such data as the basis for graphical output. (Assume that a square grid of points is used). Only a general strategy need be indicated: points of possible difficulty need not be explored!

6. Draw sketches illustrating the direction of $\nabla\phi$ at points close to (*a*) a minimum (*b*) a maximum (*c*) a saddle point.

7. Derive an approximate formula for the separation (at a particular point) of successive contours defined by $\phi(\mathbf{r}) = nc$, where c is a constant and $n = 1, 2, 3 \ldots$ (Hint: the formula should contain $\nabla\phi$). Illustrate your result with the example $\phi = \mathbf{r}\cdot\mathbf{r}, c = 1$, and $n = 10, 11$.

28.6. Contours of constant intensity of infra-red radiation in astrophysical observations. (Courtesy of D. Fegan.)

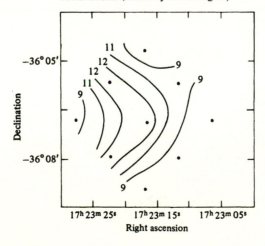

8. The electric field at $\mathbf{r} = 0$ due to a charge q at position $\mathbf{r}$ is $-q\mathbf{r}/4\pi\varepsilon_0 r^3$ (see chapter 29). If now a second, opposite, charge $-q$, is placed nearby at $\mathbf{r} + \Delta\mathbf{r}$ show that the resultant electric field of the dipole is $\mathbf{E} = \nabla(\boldsymbol{\mu} \cdot \mathbf{r}/4\pi\varepsilon_0 r^3)$ where $\boldsymbol{\mu} = q\Delta\mathbf{r}$.
   (Neglect the square and higher powers of $|\Delta\mathbf{r}|$)

9. Sketch the contours of the scalar field
   $$\phi = \tfrac{1}{2}(x^2 - y^2) + x$$
   for various (labelled) values of ϕ. Use arrows to indicate the directions and values of $\nabla\phi$. Show that the contours of ϕ intersect the contours of a second scalar field
   $$\psi = xy + y$$
   at right angles. [Hint: consider the gradient vectors $\nabla\phi, \nabla\psi$.]

10. Sketch the contours of the scalar field
    $$\phi = \cos x + \cos y$$
    for the square region $-3\pi < x, y < +3\pi$, and for the ϕ-values
    (i) $2 - \phi \ll 1$
    (ii) $\phi = 0$
    (iii) $2 + \phi \ll 1$
    Use different colours for (i), (ii), (iii).

29
The vector field

A *vector field* is a vector function of position. Such a function is used in physics to describe the force **F** on a particle, due to interaction with other particles, which contribute to the total field experienced by the particle according to such force laws as Coulomb's law. It is also basic to fluid dynamics in which the local velocity **v** of fluid flow is a function of position **r**.

In certain branches of physics and engineering the computation of vector fields offers the rewards and status of a career. For example aircraft designers need immense detail of air current/pressures and stress distributions of air frames and must compute and plot out vector fields of great complexity. General methods of systematic calculation are still the subject of mathematical research.

Let us look first at the case of a *force field*, using the particle-in-a-bowl problem of chapters 20 and 27 as our example. In the equation $d\mathbf{v}/dt = -\omega^2\mathbf{r}$, the right-hand side can be regarded as the vector function $\mathbf{F}(\mathbf{r}) = -\omega^2\mathbf{r}$. This is the force that would act on a unit mass at **r**. The time variable is absent from this expression which describes a static field. The properties of a vector field can be studied without any immediate reference to particle motion.

To express the Coulomb field due to a point charge, fixed arbitrary axes may be chosen with the charge (positive, say) at the origin, $x = y = z = 0$. The field **F** at any vector position is again parallel to **r**, with a magnitude proportional to r^{-2} where $r = (x^2 + y^2 + z^2)^{\frac{1}{2}}$ is the distance from the origin. It follows that the components are given by

$$F_x = Cxr^{-3}, F_y = Cyr^{-3}, F_z = Czr^{-3} \tag{29.1}$$

The magnitude of **F** is

$$F = (F_x^2 + F_y^2 + F_z^2)^{\frac{1}{2}} = Cr^{-2}.$$

The Coulomb field can be studied by tabulating one of its components (table 29.1), say $F_x = Cxr^{-3}$. The approximate values of F_x are calculated for the plane $y = 0$ at the points $x, z = 0, 1, 2, 3, 4, 5, 6$. For ease of subsequent graphical plotting and without regard to electrical units, the coefficient is taken as $C = 20$.

This table confirms some obvious and some less obvious features. The first row of the table shows that F_x falls off as x^{-2}. On the other hand, at a general point $(x, 0, z)$ both F_x and F_z are finite. For a fixed value z, say, $z = 5$, the table shows that F_x vanishes at $x = 0$ (where the field is parallel to z), increases in magnitude to a stationary value 0.3, then falls off, ultimately to zero.

Suppose now that two charges are present. The resultant field is then the vector sum of the separate fields due to each single charge. For distances large compared to the separation of the charges the resultant field differs only slightly from that estimated by locating the total charge at a single point. It often happens that the system is electrically neutral, that is, the charges are equal in magnitude but of opposite sign. We must make a more careful estimate of the field at large distances in this case. Such a field is described as *dipolar* and is of interest to physicists, chemists and engineers. For example simple molecules, such as CO, may be treated as dipoles for many purposes, as we have indicated in earlier chapters.

In working out the dipolar field let us take the z-axis to be the line through the two charges and the origin midway between them. Let us put the positive charge at $z = 1$ and the negative charge at $z = -1$, since the

Table 29.1 $\quad F_x(x, 0, z) = 20x(x^2 + z^2)^{-\frac{3}{2}}.$

z \ x	0	1	2	3	4	5	6
0	*	20.0	5.0	2.22	1.25	0.8	0.56
1	0	7.08	3.58	1.90	1.14	0.74	0.53
2	0	1.79	1.77	1.28	0.89	0.64	0.47
3	0	0.64	0.85	0.79	0.64	0.50	0.40
4	0	0.28	0.44	0.48	0.44	0.38	0.32
5	0	0.16	0.26	0.30	0.30	0.28	0.25
6	0	0.09	0.16	0.20	0.21	0.21	0.20

resulting field can easily be rescaled to make it more general. This arrangement provides an electric dipole directed along the positive z-axis.

All planes containing the dipole axis (z-axis) are equivalent, that is to say, the field pattern is identical in each one. There is no generality lost therefore by remaining in one such plane; let it be $y = 0$. Then the Coulomb field due to the positive charge at $(0, 0, 1)$ has zero y-component and x, z components given by

$$F_x(x, 0, z - 1) = 20x[x^2 + (z - 1)^2]^{\frac{3}{2}}$$
$$F_z(x, 0, z - 1) = 20(z - 1)[x^2 + (z - 1)^2]^{\frac{3}{2}} \tag{29.2}$$

using the expressions (29.1) for the components. To this field is added a similar one due to the negative charge at $(0, 0, -1)$ to give the resultant dipolar field, denoted by $\mathbf{D}$. For economy, since y is taken as zero, $\mathbf{D}(x, 0, z)$ is written as $\mathbf{D}(x, z)$.

$$D_x(x, z) = F_x(x, 0, z - 1) - F_x(x, 0, z + 1)$$
$$D_z(x, z) = F_z(x, 0, z + 1) - F_z(x, 0, z + 1). \tag{29.3}$$

The calculation of D_x, first of all, is quite easy – in fact all the information is contained in table 29.1, and only simple subtraction is needed. Thus to calculate $D_x(2, 3)$, select the column $x = 2$ and subtract the $z = 4$ entry (0.44) from the $z = 2$ entry (1.76), as required by (29.3), to give $D_x(2, 3) = 1.32$. Follow this procedure for all available entries.

Table 29.2 $D_x(x, z)$.

z \ x	0	1	2	3	4	5
1	0	18.22	3.24	0.94	0.36	0.16
2	0	6.44	2.15	1.12	0.50	0.24
3	0	1.50	1.32	0.80	0.46	0.26
4	0	0.48	0.57	0.48	0.34	0.22
5	0	0.20	0.28	0.28	0.23	0.17

Table 29.3 $D_z(x, z)$.

z \ x	0	1	2	3	4	5
0		−14.4	−3.58	−1.26	−0.58	−0.30
1	*	−3.58	−1.77	−0.85	−0.45	−0.26
2	17.78	5.17	0.51	−0.16	−0.19	−0.15
3	3.75	2.64	0.88	0.21	0.0	−0.04
4	1.42	1.15	0.64	0.29	0.10	0.02
5	0.69	0.61	0.42	0.24	0.12	0.15

The z-component F_z is calculated using $F_z = 20zr^{-3}$. But its calculated values are just $F_x(z, 0, x)$ i.e. table 29.1 with rows and columns interchanged. Then to find $D_x(2, 3)$ select row 2, and subtract column 4 entry (0.89) from column 2 entry (1.77) (as per formula 29.3) to give $D_z(2, 3) = 0.88$. In this way, tables 29.2 and 29.3 are constructed for the dipole field in the positive x, z quadrant.

The D_x, D_z values provided in tables 29.2 and 29.3 can be used in a variety of ways to bring out the characteristics of the dipole field and to illustrate some useful attitudes to fields in general. Following the advice given in chapter 1, the first task is to draw a figure. D_x, D_z values for each coordinate point (x, z) are represented by an arrow starting at x, z with components proportional to D_x, D_z. In fig. 29.1, the x, z points form a square grid covering the four quadrants about the dipole at the origin. Vectors are not shown for some points, their magnitude being too big to fit comfortably into the diagram.

Figure 29.1 has some obvious symmetry. First of all the left and right-hand sides are mirror images of each other. Mathematically expressed,

29.1. Dipole field. Each arrow represents the local vector field.

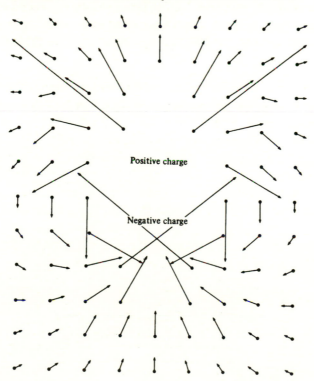

$$D_x(-x, z) = -D_x(x, z)$$
$$D_z(-x, z) = +D_z(x, z). \tag{29.4}$$

It is also apparent from fig. 29.1 that

$$D_x(x, -z) = -D_x(x, z)$$
$$D_z(x, -z) = D_z(x, z). \tag{29.5}$$

Again it follows that in a perpendicular plane bisecting the dipole, the field is parallel to the dipole, so $D_x(0, z) = 0$.

Another way of representing a vector field is to draw lines which are everywhere parallel to the local value of the vector function. These are called *streamlines* in fluid dynamics and *lines of force* in mechanics. An example is shown in fig. 29.2.

Finally, note that we have already met an important type of vector field in the last chapter – that which is defined when we operate with the gradient operator ∇ on a scalar field. If a *force* field can be expressed in this way, the corresponding scalar field is the appropriate *potential* function. Often, dynamical problems can be solved in terms of this scalar field, rather than the force field itself, by use of the principle of conservation of energy. There are great advantages to this, one of which is the avoidance of the awkward vector addition of vector fields which has caused us so much trouble here.

Given a potential function $\phi(\mathbf{r})$, the corresponding force field is

$$\mathbf{F}(\mathbf{r}) = -\nabla\phi(\mathbf{r}). \tag{29.6}$$

Given a force field can we construct a potential field, obeying (29.6)? Since integration is the inverse of differentiation, the question is: can we somehow integrate a given vector field? How to do this is considered in the next chapter, and the consequences are explored in chapter 31.

Summary

Fluid flow and electrical and gravitational forces can be expressed as vector fields. At each point in space the field has a magnitude and a direction.

The Coulomb field due to a charge at the origin is given by the vector

29.2. An example of streamlines, flow round airfoil section.

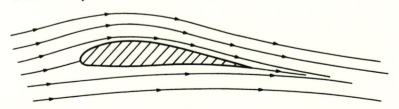

$F(x, y, z)$, where $F_x = Cxr^{-3}$, $F_y = Cyr^{-3}$, $F_z = Czr^{-3}$ $(r^2 = x^2 + y^2 + z^2)$. An electric dipole with a positive charge at $z = 1$ and a negative charge at $z = -1$ has a field in plane $y = 0$ given by $D(x, z) = F(x, 0, z - 1) - F(x, 0, z + 1)$. A vector field can often be expressed as the gradient of a scalar field, e.g. force $= -\nabla$(potential).

EXERCISES

1. *Sketch* the vector field corresponding to the 'particle-in-a-bowl' problem (*a*) as in fig. 29.1 (*b*) using lines of force.

2. *Sketch* the two-dimensional vector fields
    (*i*) $v_1 = (x, -y)$
    (*ii*) $v_2 = (y, x)$
    (*iii*) $v_1 + v_2$

3. The water in a large swimming pool is continually changed by entry and exit pipes placed at opposite ends of the pool at mid-depth. Assuming a steady flow represented by a vector field, sketch the pattern to be expected in the horizontal plane containing the pipes.

4. Describe the field given by $v = \omega \times r$.

5. A parallel plate capacitor consists of two plates which carry equal and opposite charges. These may be considered as made up of a large number of dipoles, placed side-by-side. *Sketch* the electric field within and around the capacitor.

6. Explain and summarise the symmetry of the dipole vector field (use fig. 29.1). Use it (together with the tables) to specify $D(-4, 1)$, $D(2, -3)$ and $D(-5, -5)$.

7. Explain why it would be inappropriate to use the dipolar field for the CO_2 molecule (chapter 7). *Sketch* the electric field corresponding to electric charges arranged in the same way as the atoms of this molecule, assuming overall electrical neutrality.

8. Describe the variation of the dipole field at a fixed distance from the dipole axis, say $x = 4$.

9. Discuss the correctness (or otherwise) of the following: (*a*) in steady flow a small element of fluid follows a single streamline; (*b*) in a given force field a particle follows a single line of force.

10. Write the dipolar field for two charges $\pm e$ at $(0, 0, \pm d)$. Find the form of this field in the limit $d \to 0$, $2ed = $ constant. This quantity is the magnitude of the 'dipole moment' – see chapters 3, 8, 9. This gives a very useful approximation to the field of a dipole at large distances. (see chapter 28, exercise 8.)

30
Line integration

Acceleration is defined as the rate of change of velocity. Very well then. A moving particle is decelerated by a constant force until brought to rest; by what fraction is its speed reduced at half the stopping distance? The well-drilled physics student knows that there is a formula relating speed to distance and has no difficulty with the answer, $1/\sqrt{2} = 0.71$. Suppose the constant force is now replaced by one obeying Hooke's law. The first trick no longer works, so try another – conservation of energy? With luck the answer eventually comes out, $\sqrt{3}/2 = 0.86$. But stiffen the problem further, with force varying as distance squared say, and the success rate falls rapidly.

The difficulty is that acceleration dv/dt is understood as a function of time whereas the given force F depends upon position x. In the equation of motion

$$m\frac{dv}{dt} = F(x) \tag{30.1}$$

certainly v is given by the time integral of $F(x)$, but $x(t)$ is not known until the equation has been solved for v! The way through this knot is to realise that v, like F, can be regarded as a function of x. Its time-derivative is given by the function-of-a-function rule (chapter 16),

$$\frac{dv}{dt} = \frac{dv}{dx}\frac{dx}{dt} = v\frac{dv}{dx}.$$

Using the product rule (same chapter) this can be usefully expressed as follows:

$$\frac{dv}{dt} = \frac{d}{dx}(\tfrac{1}{2}v^2).$$

Using this form (30.1) becomes a simple first-order differential equation in x for the velocity $v(x)$, in the form

$$\frac{d}{dx}(\tfrac{1}{2}mv^2) = F(x).$$

The solution for $\tfrac{1}{2}mv^2$ is the indefinite integral $\int F(x)dx$. There is a constant of integration v_1 which can be introduced by writing

$$\tfrac{1}{2}mv^2 = \tfrac{1}{2}mv_1^2 + \int_{x_1}^{x} F(u)du \tag{30.2}$$

where now, at the point $x = x_1$, the initial value is $v = v_1$. If $x(t)$ is needed a further equation must be solved, $dx/dt = v(x)$. For example, suppose $F(x) = -qx^2$ (decelerative force varying as distance squared.) Then taking $x_1 = 0$ for simplicity,

$$\tfrac{1}{2}mv^2 = \tfrac{1}{2}mv_1^2 - \tfrac{1}{3}qx^3.$$

The stopping distance c is fixed by putting $v = 0$ to give $c = (\tfrac{3}{2}mv_1^2/q)^{\frac{1}{3}}$. At the half-way mark, $x = \tfrac{1}{2}c$, which gives for v,

$$\tfrac{1}{2}mv^2 = \tfrac{1}{2}mv_1^2 - \tfrac{1}{3}q[\tfrac{1}{2}(\tfrac{3}{2}mv_1^2/q)^{\frac{1}{3}}]^3 = \tfrac{7}{16}mv_1^2.$$

Thus $v/v_1 = (7/8)^{\frac{1}{2}} = 0.94$.

In the interval Δt the particle covers the distance $v\,\Delta t$ so that x-integration of F is equivalent to time-integration of Fv. This result can be extended to two or three dimensions with, however, an important new feature. Starting with the basic vector equation of motion $m\,d\mathbf{v}/dt = \mathbf{F}$, take the scalar product with $\mathbf{v}$, and use the identity $\mathbf{v}\cdot d\mathbf{v}/dt = \tfrac{1}{2}(d/dt)|\mathbf{v}|^2$. Integrate with respect to time t to give, analogous to (30.2)

$$\tfrac{1}{2}m|\mathbf{v}|^2 = \tfrac{1}{2}m|\mathbf{v}_1|^2 + \int_{t_1}^{t_2} \mathbf{F}\cdot\mathbf{v}\,dt. \tag{30.3}$$

In (30.3), $\mathbf{v}_1$ is the initial $(t = t_1)$ value of $\mathbf{v}$.

The application here is to fields which depend only upon position $\mathbf{r}$, i.e. $\mathbf{F} = \mathbf{F}(\mathbf{r})$. This therefore excludes magnetic forces which depend also upon the velocity $\mathbf{v}$. Practically speaking, $\mathbf{F}$ is either electrical or gravitational and is taken to be static, i.e. independent of time (other, of course, than through $\mathbf{r}$ itself).

Evaluation of the time-integral as it stands requires knowing the function

$\mathbf{r}(t)$, to calculate $\mathbf{F}$ and to provide $\mathbf{v} = d\mathbf{r}/dt$. As illustration consider the particle-in-bowl problem where $F(\mathbf{r}) = -\omega^2 \mathbf{r}$ for a particle of unit mass. A general solution to the equation of motion is $x = A \cos \omega t$, $y = b \cos(\omega t - \phi)$, where A, B, ϕ are constants of integration (fixing $x = A$ at $t = 0$ really uses up a fourth constant). These orbits consist of ellipses about the origin $x = y = 0$ as centre, as in fig. 30.1. Between two fixed points, P, P', there are many geometrical paths, PQP', PRP' etc., though only one with an assigned journey time. Then $dx/dt = -\omega A \sin \omega t$, $dy/dt = -\omega B \sin(\omega t - \phi)$ so that

$$\mathbf{F} \cdot \mathbf{v} = F_x \frac{dx}{dt} + F_y \frac{dy}{dt}$$

$$= \omega^3 [A^2 \cos \omega t \sin \omega t + B^2 \cos(\omega t - \phi)\sin(\omega t - \phi)].$$

Thus, integrating from t_1 to t_2 corresponding to fixed points (x_1, y_1), (x_2, y_2),

$$\int_{t_1}^{t_2} dt\, \mathbf{F} \cdot \mathbf{v} = -\tfrac{1}{2}\omega^2 A^2 [\cos^2 \omega t_2 - \cos^2 \omega t_1]$$

$$-\tfrac{1}{2}\omega^2 B^2 [\cos^2(\omega t_2 - \phi) - \cos^2(\omega t_1 - \phi)]$$

$$= -\tfrac{1}{2}\omega^2 (x_2^2 + y_2^2) + \tfrac{1}{2}(x_1^2 + y_1^2)$$

$$= -\tfrac{1}{2}\omega^2 [r_2^2 - r_1^2]. \tag{30.4}$$

The time-integral therefore depends only upon the initial and final positions.

   Less obvious is the idea of a purely geometric evaluation without any reference to dynamical paths. Figure 30.2 shows the point (x_1, y_1) connected by three different paths to (x_2, y_2). The first is made up of the straight line sections $(x_1, y_1) \rightarrow (x_2, y_1) \rightarrow (x_2, y_2)$. Such a path cannot

30.1. Elliptical paths from P to P'

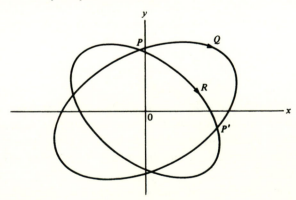

correspond to any orbital solution of (30.2). Along the first leg, $v_y = dy/dt = 0$, and the integral becomes

$$\int_{t_1}^{t_2} F_x \frac{dx}{dt} dt = \int_{x_1}^{x_2} F_x dx.$$

In this re-expression it looks as though the dt has been 'cancelled'. More precisely there is a change of independent variable from t to x, making use of the function-of-a-function rule (chapter 13). Since $F_x = -\omega^2 x$, the x-integration from x_1 to x_2 gives the result, $-\frac{1}{2}\omega^2(x_2^2 - x_1^2)$. Quite similarly, along the second section $(x_2, y_1) \rightarrow (x_2, y_2)$, x remains constant and the y-integration gives

$$\int_{y_1}^{y_2} F_y dy = \int_{y_1}^{y_2} (-\omega^2 y) dy = -\frac{1}{2}\omega^2(y_2^2 - y_1^2).$$

The alternative route $(x_1, y_1) \rightarrow (x_1, y_2) \rightarrow (x_2, y_2)$ gives the same contributions but in the opposite order. Either way, the sum of the contributions is the same as (30.4).

The kind of integral with which we have been dealing can be rewritten as

$$\int_{t_1}^{t_2} \mathbf{F} \cdot \mathbf{v} \, dt = \int_{x_1, y_1}^{x_2, y_2} (F_x \, dx + F_y \, dy)$$

$$= \int_{\mathbf{r}_1}^{\mathbf{r}_2} \mathbf{F} \cdot d\mathbf{r}. \tag{30.5}$$

The right-hand side of (30.5) is described as a vector integral or *line integral*. The notation which uses $d\mathbf{r}$ to represent (dx, dy, dz) is neat but can be confusing. Remember that such an integral is always taken over some path, so that we can trace out this path with a function $\mathbf{r}(t)$ and reverse the above steps in order to express the integral as the left-hand side of (30.5), which has a more obvious meaning.

30.2. Alternative paths from P to P'.

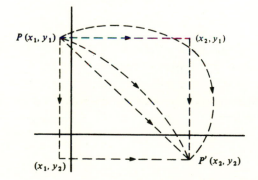

The independence of the vector integral upon geometrical path which we have found is *not* a general property of vector fields but is a very useful one whenever it is found. It can be expressed in another way. Starting at the point (x_1, y_1) in fig. 30.1, an integral could be taken round a closed loop, for example, $(x_1, y_1) \rightarrow (x_2, y_1) \rightarrow (x_2, y_2) \rightarrow (x_1, y_2) \rightarrow (x_1, y_1)$. In tracing the latter two steps, the x and y integrations which we have already evaluated are reversed in sign because quite generally 'integral from B to A' is the negative of 'integral from A to B'. But since the integrals associated with the two routes in the original order were equal, it follows that the integral around the loop must add up to *zero*. The result in this case applies to *any* such path of integration and may be expressed as

$$\int_C \mathbf{F} \cdot d\mathbf{r} = 0 \qquad\qquad (30.6)$$

where the symbol C refers to any closed loop. What vector fields have this property, enabling us to define a unique integral between any two points? The full answer will not emerge until chapter 34. In the meantime let us accept (30.6) as the definition of a *conservative* vector field.

In the next chapter, we return to the question of the relation between force and potential, using the technique of line integration. We shall see that the conservative property defined here leads to the definition of the potential energy, which might already be anticipated from the form of (30.3).

Summary
The time integral of $\mathbf{F} \cdot \mathbf{v}$ along a path from $\mathbf{r}_1$ to $\mathbf{r}_2$ may be written as

$$\int_{t_1}^{t_2} \mathbf{F} \cdot \mathbf{v} \, dt$$

or

$$\int_{\mathbf{r}_1}^{\mathbf{r}_2} \mathbf{F} \cdot d\mathbf{r} = \int_{x_1, y_1, z_1}^{x_2, y_2, z_2} (F_x \, dx + F_y \, dy + F_z \, dz)$$

where (in general) the path of integration must also be specified.

If $\mathbf{F}$ stands for an electrical or gravitational field or, in general, a conservative field, the vector integral depends only upon $\mathbf{r}_1, \mathbf{r}_2$; it is independent of the path. Path independence may be described by the equation,

$$\int_C \mathbf{F} \cdot d\mathbf{r} = 0$$

where the integration path is any closed loop C.

EXERCISES

1. Draw a rough sketch of a vector field for which (30.6) would clearly not be valid.

2. For the two-dimensional field, $\mathbf{F} = (ky, kx)$
   (*i*) sketch its vector pattern (see chapter 29);
   (*ii*) Show that $\int_C \mathbf{F} \cdot \mathbf{dr} = 0$ where C denotes the circle, radius R, centred at the origin. (Hint; use polar coordinates, $x = R \cos \phi, y = R \sin \phi$).

3. For the two-dimensional vector field,
   $$E_x = (1 + y^2)^{-1}, \quad E_y = -2xy(1 + y^2)^{-2}$$
   (*i*) sketch the vector pattern;
   (*ii*) evaluate $\int \mathbf{E} \cdot \mathbf{dr}$ from $\mathbf{r} = 0$ to $x = c$, $y = c$ along the paths
   (*a*) $0 \rightarrow (c, 0) \rightarrow (c, c)$; (*b*) $0 \rightarrow (0, c) \rightarrow (c, c)$, (*c*) $0 \rightarrow (c, c)$. All segments are to be straight lines.

4. It should be clear that the vanishing of the integral (30.6) applies to the dipole field of chapter 29. Make a rough check of this, as follows. Consider the integral around a square whose corners are $(1.5, 5.5), (3.5, 4.5), (1.5, 3.5),$ $(0.5, 4.5)$. Estimate the integral using only the values of the field at the midpoints of the sides and show how the four contributions *roughly* cancel in this approximation.

5. This chapter has really only concentrated on one kind of line integral. The following illustrates another. A ring of arbitrary shape has constant mass per unit length ρ. It is placed in a gravitational force field $\mathbf{F}(\mathbf{r})$. Discuss how the total force on the ring might be expressed as a line integral.

6. In the text example relating to the field $\mathbf{F} = -\omega^2 \mathbf{r}$, evaluate $\int \mathbf{F} \cdot \mathbf{dr}$ along the direct path from (x_1, y_1) to (x_2, y_2) (the diagonal line in fig. 30.2).

7. Evaluate directly the line integral $\int_C \mathbf{E} \cdot \mathbf{dr}$, where $\mathbf{E} = (f_1 x, f_2 y)$, (with f_1, f_2 constant), and where C denotes the positive quadrant of a circle, centre at the origin, of radius R.

31
The potential field

When it comes to energy, conservation has a stricter meaning for the physicist than for the ecologist. A freely-falling object gains kinetic energy, and the gain depends only upon the loss of height. So height then provides available or potential energy which can be drawn upon at every point. It is treated as a function of the vertical coordinate and can be studied mathematically without any reference to motion. Unlike the gravitational force to which it is related, it is not a directed quantity. For this reason the potential function is described as a *scalar field* (chapter 28).

The mathematics needed to express these notions stems from formulae (30.3), (30.5) and (30.6). Combining the first two,

$$\tfrac{1}{2}m|\mathbf{v}|^2 - \int_{\mathbf{r}_1}^{\mathbf{r}} \mathbf{F}\cdot d\mathbf{r} = \tfrac{1}{2}m|\mathbf{v}_1|^2 \qquad (31.1)$$

where $\mathbf{r}_1$ is the initial vector position and $\mathbf{r}$ the final position. The constant gravitational force is described by the field $\mathbf{F} = -mg\mathbf{k}$, where $\mathbf{k}$ is the unit vector in the positive (upward) vertical direction. Thus

$$\int_{\mathbf{r}_1}^{\mathbf{r}} \mathbf{F}\cdot d\mathbf{r} = -mg \int_{z_1}^{z} dz.$$

The integration can be taken along any geometrical path (chapter 30) gravitational force is described by the field $F = -mg\mathbf{k}$, where $\mathbf{k}$ is the unit joining initial point P at $\mathbf{r}_1 = (x_1, y_1, z_1)$ to $\mathbf{r} = (x, y, z)$. Three possible paths are illustrated schematically in fig. 31.1. The final points Q, Q', Q'' etc., depend upon the initial velocity at P.

The integration gives simply $-mg(z - z_1)$, regardless of the initial and final horizontal coordinates. Equation (31.1) then becomes

$$\tfrac{1}{2}m|\mathbf{v}|^2 + mgz = E \tag{31.2}$$

where E is the initial value of the left-hand side, namely $E = 1/2m|\mathbf{v}_1|^2 + mgz_1$.

In this example, the energy mgz is regarded as the *potential field*, and written

$$\phi(z) = mgz.$$

Added to kinetic energy it provides the total energy E, which according to (31.2) remains constant, i.e. is conserved.

We now see how various definitions fit together. If a force field is *conservative*, i.e. satisfies (30.6), then a corresponding potential field can be defined according to

$$\phi = -\int_{\mathbf{r}_1}^{\mathbf{r}} \mathbf{F} \cdot d\mathbf{r}. \tag{31.3}$$

We can obtain $\mathbf{F}$ from ϕ by the inverse relation (29.6). (The proof of this 'Fundamental Theorem of Calculus' for vectors is studied in exercise 4.)

Let us look at a more difficult case. For stone throwing (31.2) may suffice, but for missiles, satellites and planetary motion it is necessary to use the inverse square law of gravitational attraction. Mathematically this has the same form as the electrical (Coulomb) law so that the two instances may be treated together.

31.1. Trajectories of a particle in a uniform gravitational field.

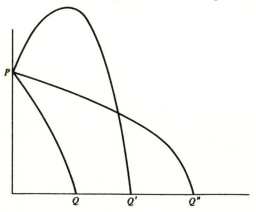

The inverse square law is expressed through the force field $\mathbf{F} = C\mathbf{r}/r^3$, where the point charge or mass centre is at the origin $\mathbf{r} = 0$. The coefficient C depends upon units and for mathematical purposes can be given any convenient numerical value. What is needed is again the line-integral $\int_{\mathbf{r}_1}^{\mathbf{r}} \mathbf{F}\cdot d\mathbf{r}$. For the time being we shall assume that this is independent of path; the route from $\mathbf{r}_1$ to $\mathbf{r}$ can be conveniently taken as PRQ, as shown in fig. 31.2. This important assumption will be vindicated in chapter 34 for the inverse square law potential.

Referring to fig. 31.2, the position vector of the initial point P is $\overrightarrow{OP} = \mathbf{r}_1$. In the first (circular) section PR, $|\mathbf{r}|$ is constant and equal to $|\mathbf{r}_1|$, thus $\int_P^R \mathbf{F}\cdot d\mathbf{r} = 0$, since $\mathbf{F}$ is parallel to $\mathbf{r}$ which is perpendicular to the path element. The connecting section RQ is parallel to $\mathbf{r}$ so that

$$\int_R^Q \mathbf{F}\cdot d\mathbf{r} = C\int_{r_1}^r r^{-2}\,dr$$
$$= C(r_1^{-1} - r^{-1}).$$

Equation (27.1) becomes

$$\tfrac{1}{2}mv^2 + C/r = E$$

where r denotes $|\mathbf{r}|$ and E is the initial value of the left hand side, namely $\tfrac{1}{2}m|\mathbf{v}|^2 + Cr_1^{-1}$. Comparing with (31.2), the inverse square law force $\mathbf{F} = C\mathbf{r}/r^3$ evidently corresponds to the potential field.

$$\phi(x, y, z) = C/r = C(x^2 + y^2 + z^2)^{-\frac{1}{2}}. \tag{31.4}$$

As a final illustration the dipole potential is calculated. Since the dipole field (chapter 29) is the sum of two Coulomb (inverse square) fields, its corresponding potential is the sum of two corresponding potentials. Taking

31.2. Integration path for the evaluation of $\int_{\mathbf{r}_1}^{\mathbf{r}_2} \mathbf{F}\cdot d\mathbf{r}$ for the Coulomb potential.

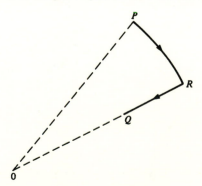

as before the positive and negative charges at $z = 1$, and $z = -1$ respectively,

$$\phi(x, y, z) = C[x^2 + y^2 + (z - 1)^2]^{-\frac{1}{2}}$$
$$- C[x^2 + y^2 + (z + 1)^2]^{-\frac{1}{2}}. \quad (31.5)$$

As an alternative to the procedure used in chapter 29, we could take the gradient of this to obtain the force field, which is much easier.

Summary

The potential field $\phi(\mathbf{r})$ is the potential energy of a particle and is a function of position vector $\mathbf{r}$.

The potential field corresponding to gravitation (close to the Earth's surface) is $\phi = mgz$ where z is the vertical coordinate measured from some arbitrary level.

The potential field corresponding to the inverse square force is of the form $\phi = C/r$, where r is the distance of the field point to the charge or mass centre. The coefficient C is positive for a repulsive force, and negative for an attractive force.

These are special cases of the general definition

$$\phi = -\int_{\mathbf{r}_1}^{\mathbf{r}} \mathbf{F} \cdot d\mathbf{r}$$

which applies to any conservative field, and is equivalent to $\mathbf{F} = -\nabla\phi$. The dipolar field may be obtained by combining the potential field of positive and negative charges.

EXERCISES

1. A two-dimensional force field is given as $\mathbf{F} = C(x, -y)$. Can it be represented by a potential field?

2. In two dimensions, a potential function $\phi(x, y)$ is given by
   $\phi(x, y) = x^4 - Bx^2y^2 + y^4$.
   Determine the value of B required to make ϕ satisfy the differential equation,
   $$\frac{\partial^2\phi}{\partial x^2} + \frac{\partial^2\phi}{\partial y^2} = 0.$$

3. Use the approximations of chapter 15 to show that the dipole potential, equation (31.5), is given approximately by
   $\phi = 2C \cos\theta/r^2$
   for $r \gg 1$. (Cf. chapter 28, exercise 8, chapter 29, exercise 10.)

4. Supply a proof of the Fundamental Theorem discussed in the text. One

approach: first prove it for the special case of constant **F** and **r**(t) a straight line, using (28.2), then argue that the general case can be reduced to this.

5. A pair of electric coils placed on a common axis but carrying opposite currents produce a magnetic field in the gap separating them. It is modelled by the magnetostatic potential
$$\phi = B_0 x - \tfrac{1}{2}\alpha x(y^2 + z^2)$$
where x is the axial direction, and the origin is at the midpoint of the gap. Evaluate the magnetic field **B** $= -\nabla\phi$ and sketch its lines of force (B_0 and α are constants).

6. When a conducting sphere is placed in a uniform electric field, charge is redistributed on its surface until there is no longer any tangential component of the total field there. (Only then can there be equilibrium, i.e. zero current in the sphere.) Make a careful sketch of the equipotentials and field lines (lines of force) for this case.

32
Surface and volume integration

We have seen how some of the problems of handling vector fields may be avoided with the help of line integrals to construct corresponding potential fields. It turns out that integrals over surfaces are also very useful. These often take the form

$$\Phi = \int_S \mathbf{F} \cdot d\mathbf{S} \tag{32.1}$$

where S is a closed surface.

As with line integrals, the meaning of the *surface integral* in (32.1) is not immediately obvious. If we approximate the surface by a lot of small planar elements of area ΔA, it represents the sum of $\mathbf{F} \cdot \mathbf{n} \Delta A$, where $\mathbf{n}$ is the outwardly directed normal to the surface, in the limit in which the size of the elements goes to zero (as in fig. 4.5). Hence an alternative notation is

$$\Phi = \int_S \mathbf{F} \cdot \mathbf{n} \, dA \tag{32.1a}$$

and this can in turn, be reduced to a double integral for evaluation, as we shall see later. Such a surface integral can also be taken over a surface which is not closed but in that case we must be careful to identify the direction of $d\mathbf{S}$.

Physicists often try to avoid having to wrestle with double integrals, by choosing surfaces for which the value of (32.1) is more or less obvious, whenever they are forced to introduce surface integrals. In particular, it is sometimes possible to pick a surface for which the quantity $\mathbf{F} \cdot \mathbf{n}$ is simply a constant.

Such integrals are particularly useful in the case of electric, gravitational or other inverse square law force fields. Gauss's law in electrostatics states

$$\varepsilon_0 \int_S \mathbf{E(r)} \cdot d\mathbf{S} = \text{enclosed charge} \tag{32.2}$$

for any closed surface S, where $\varepsilon_0 \approx 8.8 \times 10^{-12}$ coulombs/volt metre, and analogous rules hold for the other inverse square law potentials. We shall not prove this here – it must await further developments in chapter 34. However, note that it works for the most obvious case – a spherical surface of radius R surrounding a charge of magnitude Q. In this case $\mathbf{E} \cdot \mathbf{n}$ is constant on the surface, so we can write

$$\varepsilon_0 \int_S \mathbf{E} \cdot d\mathbf{S} = \varepsilon_0 \times \text{integrand} \times \text{area}$$
$$= \varepsilon_0 \times (Q/4\pi\varepsilon_0 R^2) \times 4\pi R^2$$
$$= Q$$

as (32.2) stipulates.

Gauss's law may also be simply applied to a uniform infinite planar sheet of charge, such that the charge per unit area is q. In this case, a 'pill box' surface is appropriate, as shown in fig. 32.1. It is evident from symmetry that $\mathbf{E}$ must everywhere be directed normal to the sheet of charge, so it is parallel to the sides of the pill box and makes no contribution to (32.2) there. Symmetry also demands that $\mathbf{E} \cdot \mathbf{n} (= E_n)$ be uniform and equal in magnitude on the two end faces, so that Gauss's law gives

$$2\varepsilon_0 A E_n = Aq,$$

from which the area A may be cancelled to obtain

$$E_n = q/2\varepsilon_0. \tag{32.3}$$

32.1. Cylindrical 'pill-box' surface used in applying Gauss's law to a sheet of charge intersecting the pill-box as shown by the dotted line.

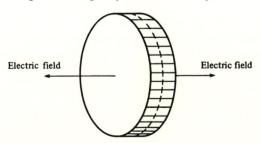

Electric field ←   → Electric field

Now suppose we place two such sheets with opposite changes, side by side, separated by a distance d, as in fig. 32.2. We can use (32.3) to get the field due to each sheet of charge separately and add them. The two contributions combine to give a field

$$E = q/\varepsilon_0 \tag{32.4}$$

between the two sheets, zero outside. Thus we have neatly solved the 'parallel-plate capacitor' problem in electrostatics (cf. exercise 5, chapter 29).

It is instructive (and a cautionary tale) to analyse this problem without Gauss's law. To do so, we can consider the capacitor to be a uniform array of dipoles, and add up their contributions to the field. We shall take $d = 1$ for the purposes of this calculation.

Charge associated with an element of area at (x, y) on the positively charged plane provides a z-component of the field at $(0, 0, z)$ given by

$$C \frac{(z-1)}{[x^2 + y^2 + (z-1)^2]^{\frac{3}{2}}}$$

(See chapter 29). The coefficient C used previously is equal to the amount of charge (in coulombs) multiplied by $(4\pi\varepsilon_0)^{-1}$.

The total field is now found by adding the contributions from all such charges. Suppose the charge per unit area is q, which in practice means Q/A where Q is the total charge and A is the area of a very large plate. Then the amount of charge in a small area of dimensions Δx, Δy is the product $q\Delta A = q\Delta x\Delta y$. In the infinitesimal limit, the field sum due to the charges distributed over all the positive plane is given by the surface integral (expressed as a *double integral*):

$$(4\pi\varepsilon_0)^{-1} q \int_{-\infty}^{\infty} dy \int_{-\infty}^{\infty} dx (z-1)$$
$$\times [x^2 + y^2 + (z-1)^2]^{-\frac{3}{2}}. \tag{32.5}$$

32.2. Parallel-plate capacitor.

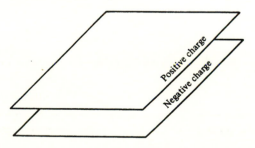

Positive charge

Negative charge

This kind of reduction to a double integral can be performed on surface integrals over more complicated surfaces as well but is quite a tricky exercise in such cases. Since the sheet is supposed to be limitless, i.e. has no edges, the x and y integrations (the order does not matter) are over all positive and negative values. Evaluation is not as difficult as it looks. A change of variables is recommended. Write $x = r\cos\phi$, $y = r\sin\phi$, where r and ϕ are polar coordinates. The geometrical meaning is that ϕ is the angle made by the radius vector $\mathbf{r}$ with the x-axis. So $x^2 + y^2 = r^2$, independent of ϕ – but what about the infinitesimal area element $dxdy$? Figure 32.3 shows the plane divided up by lines of constant x, y and r. Replacing the area element $dxdy$ by $r\,d\phi\,dr$ the double integral becomes

$$\int_0^\infty dr \int_0^{2\pi} d\phi \frac{(z-1)r}{[(z-1)^2 + r^2]^{\frac{3}{2}}},$$

where the range of integration for ϕ is 0 to 2π, while the radius r can vary from 0 to ∞. This is another double integral, but a simpler one.

Since the field is independent of ϕ, $\int_0^{2\pi} d\phi$ merely contributes a factor 2π, and the integral becomes

$$\pi \int_0^\infty \frac{2r\,dr}{[(z-1)^2 + r^2]^{\frac{3}{2}}}.$$

Now the numerator $2r\,dr$ is just the derivative of $(z-1)^2 + r^2$ so that the indefinite integral is $-2[(z-1)^2 + r^2]^{-\frac{1}{2}}$. At the upper limit this function vanishes and at $r = 0$ it is $-2(z-1)^{-1}$. Thus the $(z-1)$ factors cancel out

32.3. Cartesian and polar area elements.

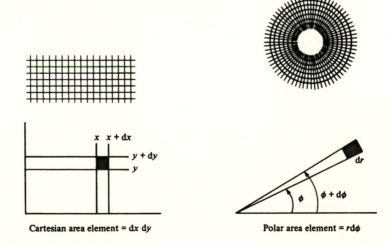

Cartesian area element = $dx\,dy$ Polar area element = $r\,d\phi$

and the $z = 1$ plane contribution to the field is just $q/2\varepsilon_0$. By symmetry the $z = -1$ negative charge gives exactly the same contribution so that finally the net interior constant field is q/ε_0, as before. Thus a page of careful mathematics has indeed led us to the same result as did Gauss's Law in a few lines...

More generally, if we seek to evaluate (32.2) for continuous distributions of charge, we may be faced by *volume integrals*, of the form

$$\int_V \rho(\mathbf{r})\,\mathrm{d}V.$$

Again, we shall not be concerned with the manipulation of such an integral here, merely with its meaning, which is the value of $\Sigma \rho \Delta V$ in the limit of small ΔV, these being elements of volume making up the total volume V, and containing charge distributed according to the charge density $\rho(\mathbf{r})$.

In principle such integrals are reducible to triple integrals (e.g. $\mathrm{d}V \rightarrow \mathrm{d}x\,\mathrm{d}y\,\mathrm{d}z$). In practice, they can be very awkward in cases where the integral does not take a simple form or the bounding surface has a complicated shape. In such cases, numerical techniques may be employed, but even these can be quite taxing. The Monte Carlo method which we mentioned briefly in chapter 17 has much to recommend it. The integral is represented by a sum over randomly chosen points, with zero contribution from those which lie outside the surface. Provided there is a simple test for the latter, this gives an easy method of evaluation, if rather inefficient.

Summary

Gauss's law in electrostatics

$$\varepsilon_0 \int_S \mathbf{E} \cdot \mathrm{d}\mathbf{S} = \text{enclosed charge}$$

involves a surface integral, which can be rewritten using $\mathrm{d}\mathbf{S} = n\,\mathrm{d}A$ where $\mathrm{d}A$ is an element of area on the surface S.

The right-hand side is a volume integral

$$Q = \int_V \rho(r)\mathrm{d}V.$$

EXERCISES

1. Electrical charge, q per unit length, is distributed along a very long line. What is the direction of the resultant electrical field $\mathbf{E}$ at any (off-axis) point? Use Gauss's law to evaluate the magnitude of $\mathbf{E}$ at a distance R from the line.

2. Using Gauss's law, evaluate $\int \mathbf{E} \cdot d\mathbf{S}$ over the surface of a sphere, radius R, for the following charge distributions:

   (*i*) charge Q centred at the centre;

   (*ii*) dipole anywhere inside the sphere;

   (*iii*) charge Q uniformly distributed over an inner concentric sphere, radius r.

   In the last case use Gauss's Law to also demonstrate that $\mathbf{E} = 0$ *inside* the sphere.

3. Write a computer program which will generate N random points on a sphere of radius R. Use it to test Gauss's Theorem for a charge placed in a sphere, but not at its centre, by approximating the integral in (32.2) as

$$4\pi R^2 N^{-1} \sum_{V=1}^{N} \mathbf{n}_i \cdot \mathbf{E}_i$$

   in the spirit of the Monte Carlo method. Optional: examine the same problem analytically.

4. In fluids, the normal force per unit area acting on a surface is given by the local pressure. How would you write the integral which gives the total force due to pressure on a closed surface within the fluid? This is a slightly different type of surface integral from that in the text. For equilibrium under gravity, this must equal the weight of fluid enclosed. Show that such a 'physical' argument leads to *Archimedes' principle*, and avoids tedious integration in analysing the total force due to pressure on a submerged body.

5. The density of a galaxy is formed to be well approximated by $\rho(r) = C \exp(-dr^3)$ where r is distance from the centre and C and d are constants. Using Gauss's law and Newton's law for the gravitational field of a point mass ($G(m/R^2)$), work out the gravitational field at distance r_0. This may be done by analogy with the electrostatic problems discussed in the text. This will require you to find the mass within a sphere, which can be expressed as an integral over r. Rewrite this integral also as one over x, y, z – not a good choice of coordinates!

6. Calculate $\Phi = \int_S \mathbf{E} \cdot d\mathbf{S}$, where S denotes a disc area, and $\mathbf{E}$ is the electric field due to an electric charge placed directly over the disc centre. Assume that the disc subtends cone angle of $2\theta_0$ at the charge.

7. Assuming that the radiation flux $\mathbf{W}$ due to the pole star is constant in magnitude and direction, calculate the value of $\Phi = \int_S \mathbf{W} \cdot d\mathbf{S}$ over the Earth's surface (a hemisphere of radius R).

   [Hint: Find the area between latitudes $\theta, \theta + d\theta$.]

33
Flux and divergence

In physical applications, the integral in (32.1) is often called a *flux*. The word suggests the flow of material through the surface over which the integral is defined. This is indeed precisely its meaning, if the vector field in the integral is the velocity field associated with fluid flow. Equally, in diffusion problems, it can be related to the average local velocity of particles.

The electrical use of flux was introduced in chapter 32 dealing with surface integration and Gauss's law. In this important application (along with magnetism) the notion of flow is to be understood by analogy, since there is no material transfer.

To fix the basic idea recall the problem of rain collection in chapter 5. The

33.1. The flux of rain through a frame held at an angle θ to the direction of the rain is proportional to $\cos \theta$.

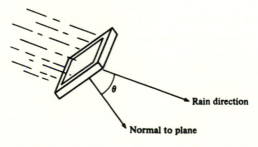

intensity of rain driven by wind could be measured by the mass of water transferred per second through a frame held perpendicular to the rain's direction. If the frame is turned through an angle θ about one edge (or equivalently, if the rain changes direction by the same angle) the effective aperture is reduced by the factor $\cos\theta$. The flux of water J (measured in kg s^{-1}) through the inclined frame becomes $J\cos\theta$. For $\theta = 90°$ the intercepted flux vanishes (see fig. 33.1).

Returning to our flux integral (32.1), we see that this is the appropriate generalisation to a surface which is not necessarily planar. Nor need it be closed, as in the discussion of chapter 32. But it often is closed because its applications in physics are generally related to conservation laws, which relate the rate of build-up of some quantity within a closed surface to a flux integral over the surface. We derived the one-dimensional diffusion equation in just this way in chapter 25 – our concern now is to do this sort of thing in three dimensions.

For example, if a fluid is incompressible, the mass of fluid within any closed surface remains constant, hence the total flux of velocity **v** over any closed surface is zero. The physical notion of *incompressibility* thus finds a mathematical expression. But there is a much neater statement which is equivalent to this, as we shall see. This is because the flux over any closed surface can generally be replaced by the corresponding *volume* integral of a related quantity, according to the Divergence Theorem of Gauss,

$$\int_S \mathbf{v}\cdot\mathbf{dS} = \int_V \text{div}\,\mathbf{v}\,dV. \tag{33.1}$$

The integrand on the right-hand side is the scalar field defined by the *divergence* of **v**,

$$\text{div}\,\mathbf{v} = \frac{\partial v_x}{\partial x} + \frac{\partial v_y}{\partial y} + \frac{\partial v_z}{\partial z} \tag{33.2}$$

or, using the notation which we mentioned speculatively in chapter 28,

$$\text{div}\,\mathbf{v} = \mathbf{\nabla}\cdot\mathbf{v} \tag{33.3}$$

It is no coincidence that the name of Gauss occurs here and in connection with (32.1) – they are really mathematical and physical versions of the same equation!

If for the moment we accept the theorem, we can give another definition of the divergence at any point as

$$\text{div}\,\mathbf{v} = \lim_{V\to 0}\left[V^{-1}\int_S \mathbf{v}\cdot dS \right] \tag{33.4}$$

where the limit implies that the surface S is shrunk to zero around the point in question, and V is the corresponding volume. The divergence tells us whether **v** is, on average, diverging from or converging to the point, so the name is sensible enough.

Proving the theorem really amounts to showing that, if div **v** is defined by (33.4), this is equivalent to (33.2) since, given this, we can break a large volume into small ones and proceed in an obvious way. Let us therefore examine (33.4) for a small cube of side ε, as in fig. 33.2.

The fluxes associated with the various faces of the cube are easily written down and evaluated in the limit $\varepsilon \to 0$. We obtain

$$\lim_{V \to 0} V^{-1} \int \mathbf{v} \cdot \mathbf{dS} = \lim_{\varepsilon \to 0} \varepsilon^{-3} \times \varepsilon^2 \times (v_x(x + \varepsilon, y, z) - v_x(x, y, z))$$

$+$ similar terms for other faces of the cube

$$= \frac{\partial v_x}{\partial x} + \frac{\partial v_y}{\partial y} + \frac{\partial v_z}{\partial z}. \tag{33.5}$$

Using a cubic element of volume in this way is not quite sufficient to prove the Divergence Theorem, since an arbitrary volume cannot be cut into cubes, but the proof can be generalised to distorted cubes easily, and that is enough. The above will suffice to convey the spirit of this type of proof and some feeling for the meaning of div **v**. For a rigorous proof, and proper restrictions on the functions involved, calculus textbooks should be consulted. In these, the Divergence Theorem often emerges in one line as a special case of Green's Theorem, with much gain in formal elegance and some loss of insight.

Let us now return to the incompressible fluid and give the promised

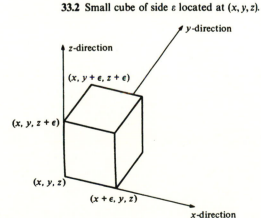

33.2 Small cube of side ε located at (x, y, z).

reformulation of the mathematical expression of incompressibility. This can now be stated as

$$\text{div } \mathbf{v} = 0 \qquad\qquad (33.6)$$

and this is indeed the basic equation of that branch of fluid dynamics which deals with incompressible fluids.

Figure 33.3 shows a simple example of fluid flow which obeys (33.6) everywhere except at the origin.

It is

$$\mathbf{v} = C\mathbf{r}/r^3 \qquad\qquad (33.7)$$

where C is a constant.

You should check by differentiation that the above statement is correct, i.e. that (33.6) holds for $\mathbf{r} \neq 0$. What would such a field mean, in terms of fluid flow? Clearly the fluid is flowing away from the origin in a manner consistent with incompressibility, but there is a point *source* of fluid at the origin. We might imagine that a thin pipe leads to this point and the fluid is being pumped down it. If, instead, it was being sucked up (33.7) would have the opposite sign and we would speak of a *sink*. In both cases div $\mathbf{v}$ is infinite at the origin. This poses some difficulties for the application of the Divergence Theorem, but it is easy to see what happens.

With a little thought we can see that the flux through any surface enclosing the origin will be the same finite quantity,

$$\int_S \mathbf{v} \cdot \mathbf{dS} = 4\pi C. \qquad\qquad (33.8)$$

The right-hand side is obtained by using, for example, the unit sphere. The divergence associated with a source or sink is really the three-dimensional

33.3. Fluid flow from a point source.

equivalent of the 'delta function' which we have studied in chapter 25.

In the presence of various sources and sinks, an appropriate generalisation of (33.8) must be written down and the flux over a surface is related to whatever sources/sinks it encloses. The result is formally identical to Gauss's Theorem in electrostatics, introduced in the last chapter. Point charges may be thought of as sources/sinks of the electric (vector) field **E**. This is because in empty space div **E** = 0, which is one of Maxwell's Equations – really it is just a consequence of the Coulomb's law (compare 33.7!) and the vector addition law. There is thus a close mathematical connection between incompressible fluid flow and electrostatics, often obscured by their different physical contexts.

Finally, let us recall that often a vector field can be expressed in terms of the gradient of a scalar field ϕ. In that case, the vanishing of divergence is expressed as

$$\mathbf{\nabla \cdot \nabla} \phi = 0 \tag{33.9}$$

where

$$\mathbf{\nabla \cdot \nabla} = \frac{\partial^2}{\partial x^2} + \frac{\partial^2}{\partial y^2} + \frac{\partial^2}{\partial z^2}.$$

This is the Laplacian operator, more commonly written as ∇^2. Equation (33.9) is *Laplace's equation*. It applies, for example, to the electrostatic potential in free space. The classic problem of electrostatics is: for given charges (sources, sinks of $\mathbf{E} = -\nabla\phi$) solve (33.9). But ∇^2 also crops up in the three-dimensional versions of the diffusion equation, wave equation, Schrödinger's equation. When physicists dream, they dream of $\nabla^2\phi$.

Summary

The flux $\int_S \mathbf{v \cdot dS}$ of a vector field over any closed surface S is equal to the corresponding volume integral of div **v** (Divergence Theorem). where

$$\text{div } \mathbf{v} = \mathbf{\nabla \cdot v} = \frac{\partial v_x}{\partial x} + \frac{\partial v_y}{\partial y} + \frac{\partial v_z}{\partial z}.$$

For fluid flow, incompressibility implies div **v** = 0. In electrostatics the same equation (for **E**) is valid wherever there is no charge. At point sources and sinks of **v** (point charges in electrostatics) the divergence is infinite but the integral of the flux from the point is finite.

Laplacian operator:

$$\nabla^2 = \mathbf{\nabla \cdot \nabla} = \frac{\partial^2}{\partial x^2} + \frac{\partial^2}{\partial y^2} + \frac{\partial^2}{\partial z^2}.$$

EXERCISES

1. For the vector field defined by
   $$\phi = \mathbf{r}/r^4$$
   find the magnitude of the divergence at any point,
   (a) by direct evaluation,
   (b) by use of the Divergence Theorem of Gauss (and symmetry arguments).

2. Two narrow pipes are inserted in an incompressible fluid. Fluid is pumped into one and out of the other at the same rate. What pattern of streamlines would you expect?

3. In earlier chapters we saw that a rigidly rotating body (or fluid) is described by the equation $\mathbf{v} = \boldsymbol{\omega} \times \mathbf{r}$. We can now think of this as a vector field. Show that its divergence is zero, and interpret this.

4. An experiment establishes that the gravitational *force* at a point in a spherical galaxy varies as $\exp(-\alpha r)$ where r is radial distance from the centre. What is the form of the corresponding *mass* distribution?

5. Given accurate values of $\mathbf{F}$ on a regular (square) grid of points, what formula would you use to make an estimate of div $\mathbf{F}$ at the same points?

6. Starting from Gauss's Theorem for the vector field $\mathbf{F}$,
   $$\int_S \mathbf{F} \cdot d\mathbf{S} = \int_V \nabla \cdot \mathbf{F} dV,$$
   deduce the analogous relation for the scalar field u,
   $$\int_S u \, d\mathbf{S} = \int_V \nabla u \, dV.$$
   (Take $\mathbf{F}_1 = u\mathbf{i}, \mathbf{F}_2 = u\mathbf{j}, \mathbf{F}_3 = u\mathbf{k}$, where $\mathbf{i}, \mathbf{j}, \mathbf{k}$ are the unit vectors, and combine the 3 equations.) (Cf. chapter 32, exercise 4.)

   By taking u as the hydrostatic pressure p at depth z, namely $p = \rho g z$, prove Archimedes' principle, namely that the upthrust on a submerged body equals the weight of the fluid displaced.

7. The electric field $\mathbf{E}$ due to a continuous distribution of polarisable material within a volume V is given by
   $$\mathbf{E} = \int_V \nabla(\mathbf{P} \cdot \mathbf{r}/4\pi\varepsilon_0 r^3) \, dV$$
   where $\mathbf{r}$ denotes the vector drawn from the field point to the element dV, and $\mathbf{P} \, dV$ is the dipole moment of the element. Use exercise 6 to show that if the material is excluded from a sphere of radius R about the field point, then $\mathbf{E} = (1/3\varepsilon_0)\mathbf{P}$, independently of R. (Assume that the surface integral is negligible over the outer material surfaces.) This is known as the Lorentz field and employed to express the effective internal electric field.

34
Circulation and the curl

In the notation of chapter 30 the line integral

$$\int_{\mathbf{r}_1}^{\mathbf{r}_2} \mathbf{F} \cdot \mathbf{dr} \tag{34.1}$$

measures the work done by the static force field $\mathbf{F(r)}$ over a finite displacement from an initial vector $\mathbf{r}_1$ to a final vector position $\mathbf{r}_2$. For an electric or gravitational field the integral is quite independent of the integration path, so that according to chapter 31, we are allowed to associate $\mathbf{F(r)}$ with a scalar potential $\phi(\mathbf{r})$, namely

$$\mathbf{F} = -\nabla\phi(\mathbf{r}) \tag{34.2}$$

The value of the line integral (34.1) is then simply $\phi(\mathbf{r}_1) - \phi(\mathbf{r}_2)$ so that for a closed loop C, i.e. $\mathbf{r}_2 = \mathbf{r}_1$, the net work vanishes

$$\int_C \mathbf{F} \cdot \mathbf{dr} = 0. \tag{34.3}$$

Now the Lorentz force on a charged particle of velocity $\mathbf{v}$, due to a field $\mathbf{B}$, is proportional not to $\mathbf{B}$ but to $\mathbf{v} \times \mathbf{B}$ so that the magnetic loop integral

$$\int_C \mathbf{B} \cdot \mathbf{dr}$$

does not have the same meaning. Nevertheless it is highly significant. In fact according to Ampère's law,

$$\int_C \mathbf{B} \cdot \mathbf{dr} = \mu_0 I \qquad (34.4)$$

where the loop C encloses the total electric current I (including displacement current). If the only current which is present is a current I in a thin wire then (34.4) gives the same result for any loop C which goes around the wire. (The direction in which the integration is to be taken will become clear later.)

There is an important relation between the loop integral (34.4) (often called the circulation of $\mathbf{B}$ around C) and an integral over any surface S spanning C, expressed by

$$\int \mathbf{B} \cdot \mathbf{dr} = \int_S \operatorname{curl} \mathbf{B} \cdot \mathbf{dS} \quad \text{(Stokes' Theorem).} \qquad (34.5)$$

This is analogous to the Divergence Theorem of Gauss (chapter 33). The vector field $\operatorname{curl} \mathbf{B}$ is defined in terms of the gradient operator $\mathbf{\nabla}$ (chapter 28), together with a vector product, as

$$\operatorname{curl} \mathbf{B} = \mathbf{\nabla} \times \mathbf{B} \qquad (34.6)$$

The components of this may be written out using the vector product rule of chapter 6.

$$\operatorname{curl} \mathbf{B} = \left(\frac{\partial B_z}{\partial y} - \frac{\partial B_y}{\partial z}, \frac{\partial B_x}{\partial z} - \frac{\partial B_z}{\partial x}, \frac{\partial B_y}{\partial x} - \frac{\partial B_x}{\partial y} \right). \qquad (34.7)$$

Just as in the case of the Divergence Theorem, an understanding (and a proof) of Stokes' Theorem may be developed by breaking up the integrals into contributions from small regions. In this case we shall be concerned with small loops, so consider shrinking a single loop to such an extent that the variation of the field $\mathbf{B}$ around it is sufficiently small to be treated as a first-order change. To simplify the calculation, Cartesian coordinates will

34.1. Small square loop at (x, y, z).

$(x, y + \Delta y, z)$ $(x + \Delta x, y + \Delta y, z)$

(x, y, z) $(x + \Delta x, y, z)$

be used with a special choice of axes, namely with the z-axis perpendicular to the plane of the elementary loop.

The contributions to the line integral (34.4) can now be considered as arising from two pairs of parallel but laterally displaced paths, namely (i) $(x, y) \to (x + \Delta x, y)$, $(x + \Delta x, +\Delta y) \to (x, y + \Delta y)$, for which only B_x contributes and (ii) $(x + \Delta x, y) \to (x + \Delta x, y + \Delta y), (x, y + \Delta y) \to (x, y)$ for which only B_y contributes (fig. 34.1). Note that the loop is pursued in an anticlockwise direction. Now in (i) the net contribution is $B_x \Delta x - (B_x + \Delta y(\partial B_x/\partial y))\Delta x$ so that cancellation of the first-order terms gives a second-order quantity $-(\partial B_x/\partial y)\Delta x\Delta y$. (The other, comparable, second-order correction, $(\partial B_x/\partial x)(\Delta x)^2$ also cancels.) Note that the surviving term is due to a rate of change of B_x in a direction perpendicular to the x axis. The second pair of terms gives a corresponding contribution but of opposite sign. Adding the two expressions and noting that the area of the loop is $\Delta S = \Delta x \Delta y$, the line integral around the small element becomes

$$\left(\frac{\partial B_z}{\partial x} - \frac{\partial B_x}{\partial y}\right)\Delta S. \tag{34.8}$$

This might also be written curl $\mathbf{B} \cdot \Delta \mathbf{S}$, so we have shown that the line integral equals this quantity for a small element. We may ask: was there not a special choice of axes involved, which would somehow invalidate this general result? The answer lies in the invariance of the quantities involved with respect to rotations (because they involve scalar products of vectors) – this means that any choice of axes can be used.

How is this result for elementary loops to be related to that for large loops, to give Stokes' Theorem?

The key to this is a mesh figure, illustrated by fig. 34.2. This is a computer-generated graph with a deliberate coarse mesh size, each square providing a small loop of dimensions Δx and Δy, as considered above. It is evident at once that adding the line integrals of two adjacent loops with a common y-boundary, there is complete cancellation along that boundary, so they add to give the circulation around a loop of area $2\Delta x \Delta y$. Adding a third square (in the x-direction) then a fourth, and so on, the net line integral (allowing for cancellation along all the integral Δy boundaries) becomes that due to a rectangular row, $\Delta y \times N \Delta x$, where N is the number of elements in the row. If now an adjacent row is similarly regarded, then the effective loop consists of the two external boundaries, on the left and right, together with internal boundaries separated by $2\Delta y$, since again the line-integral contributions along the common x-direction boundary completely cancel. Continuing to add rows in this way, thereby building up the complete figure (roughly a circle in fig. 34.2), there is complete cancellation of the line integral contributions for all internal boundaries, leaving only the desired external boundary integral.

It is clear that this does not depend on the mesh being planar; any smooth surface can be approximated by a mesh of straight lines.

The quantity (34.8), summed over the mesh, gives the area integral on the other side of Stokes' Theorem.

We now return to Ampère's Theorem, which was our original motivation for re-expressing the integrals in this way. Although in our discussion of Ampère's Theorem the current was thought of as being concentrated in a thread of filament, it is really the loop flux of the current density field, denoted now by the vector $\mathbf{J}(x, y, z)$. This means

$$I = \int_S \mathbf{J} \cdot \mathbf{dS}. \tag{34.9}$$

So, if we rewrite *both* sides of (34.4) as surface integrals and note that the surface S and its perimeter C are quite arbitrary, it becomes clear that the integrals must be the same, i.e.

$$\nabla \times \mathbf{B} = \mu_0 \mathbf{J}. \tag{34.10}$$

This is the differential form of Ampère's law. It is one of Maxwell's

34.2. Mesh figure used in proving Stokes' theorem.

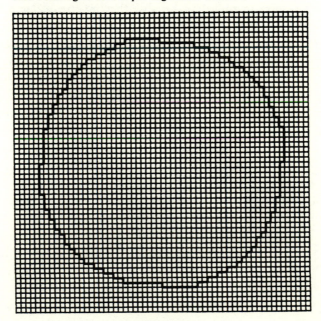

equations, the fundamental laws of electromagnetism. Given this, we could derive (34.4) by reversing our steps, and the direction of integration (or of current) will now be clear. (To keep the electromagnetic theory in order, we should emphasise that $\mathbf{J}$ is to include the displacement current; otherwise div $\mathbf{J}$ would fail to vanish identically as (34.10) requires.)

As a simple but important example consider the field

$$B_x = - k(y/r^2), \quad B_y = k(x/r^2) \tag{34.11}$$

where $r^2 = x^2 + y^2$, and k is a constant. It is easily verified that

$$\frac{\partial B_y}{\partial x} - \frac{\partial B_x}{\partial y} = 0$$

everywhere except at $x = y = 0$, where B_x, B_y are undefined. The behaviour of $\mathbf{B}$ at this singular point (axis, really) is determined by the integral condition (34.4). Thus, taking a circular loop C, of radius r, centred on the z-axis,

$$\int_C (B_x \, dx + B_y \, dy) = - (k/r^2) \int_C (y \, dx - x \, dy). \tag{34.12}$$

Following exercise 4, chapter 30, this line integral is evaluated by writing $x = r \cos \phi$, $y = r \sin \phi$ (so that $dx = - r \sin \phi \, d\phi$, $dy = r \cos \phi \, d\phi$) and integrating from $\phi = 0$ to 2π, to give simply, k, independent of r. It is concluded that $\mathbf{B}$ is generated by a filamentary current along the z-axis. Since $(\partial B_y/\partial x) - (\partial B_x/\partial y) = 0$ *except* at $x = y = 0$, the current density is zero everywhere except at the origin. In the language of the δ-function (chapter 25), it can be expressed as

$$\mu_0 J_z(x, y) = 2\pi k \, \delta(x)\delta(y). \tag{34.13}$$

To get a feel, for the general meaning of the curl of a vector field, we turn to the flow pattern of a fluid, described by the velocity field $\mathbf{v}(x, y, z)$. In the neighbourhood of interest, a reference point can be chosen, and taken (without loss of generality) as origin, $\mathbf{r} = 0$. For sufficiently small displacements $\mathbf{r}$ about the origin, each component of the vector $\mathbf{v}(\mathbf{r}) - \mathbf{v}(0)$ must consist of terms linear in x, y, z. Not all of these terms will contribute to the line integral about $\mathbf{r} = 0$ however, but only those having a certain 'skew' character. More precisely, representing their coefficients as a 3×3 matrix, it is the the antisymmetric part (see chapter 8) which contributes. Now such a 3×3 antisymmetric matrix can be written as

$$\mathbf{Mr} = \boldsymbol{\omega} \times \mathbf{r}$$

where the components of $\boldsymbol{\omega}$ provide the three finite elements of the matrix – in general depending on the point chosen. The curl of this vector is easily evaluated,

$$\text{curl}(\boldsymbol{\omega} \times \mathbf{r}) = \nabla \times (\boldsymbol{\omega} \times \mathbf{r}) = 2\boldsymbol{\omega}.$$

But the fluid flow $\boldsymbol{\omega} \times \mathbf{r}$ has a simple meaning, namely a local *rotation* of each fluid element at the angular rate $\boldsymbol{\omega}$. Remember that $\boldsymbol{\omega}$ is, in general a vector field $\boldsymbol{\omega}(\mathbf{r})$. This rotation of course is superimposed on non-rotary flow provided by $\mathbf{v}(0)$ and the symmetric part of the matrix.

Generally a field $\mathbf{F}(\mathbf{r})$ characterised by

$$\text{curl}\, \mathbf{F} = 0 \tag{34.14}$$

is described as *irrotational*, a term originating from hydrodynamics. This condition can replace the rather clumsy stipulation concerning the vanishing of line integrals (chapter 30, 31) as the necessary condition for the existence of a potential such that $\mathbf{F} = -\nabla\phi$. In magnetostatics, where $\mathbf{J} = 0$, and boundary conditions or permanent magnets replace the necessary generating currents, $\mathbf{B}$ itself can become irrotational and can accordingly be regarded as derivable from a potential – the magnetostatic potential.

Similarly, in the special case of irrotational fluid flow, the velocity field can be associated with a 'potential'.

However non-vanishing 'curl' often occurs in the theory of viscous fluid flow. In the neighbourhood of the surface there is often a tangential velocity gradient, because the tangential velocity goes to zero at the surface. Taking the surface in the (x, y) plane, the velocity components v_x, v_y depend upon the third coordinate, z. It is then evident that the loop integral $\int_C \mathbf{v} \cdot d\mathbf{r}$ taken round a rectangle with sides parallel and perpendicular to z cannot vanish. In terms of (34.14) the Cartesian component $(\text{curl}\, \mathbf{v})_z \neq 0$, so that viscous flow cannot be described in terms of the gradient of a scalar field.

A more specialised instance occurs in the description of the magnetic field and current density in superconductive materials. A tangential magnetic field external to a boundary surface of a superconductor penetrates the material but reduces with depth. This is mathematically similar to the above viscous drag and is accordingly described by a non-vanishing curl.

Summary
Stokes's Theorem:

$$\int_C \mathbf{F} \cdot d\mathbf{r} = \int_S \text{curl}\, \mathbf{F} \cdot d\mathbf{S}$$

where $\text{curl}\, \mathbf{F} = \nabla \times \mathbf{F}$

In electromagnetism this leads to

$$\mathbf{V} \times \mathbf{B} = \mu_0 \mathbf{J}$$

A vector field such that its curl is zero is irrotational. Only for an irrotational field can a corresponding scalar field, related to the field by the gradient operator (as in chapter 31), be constructed.

EXERCISES

1. Show that 'curl curl = grad div − $\mathbf{V}^2$' by writing out components. Comment on the relation between this and a basic vector identity of chapter 6, and on the dangers of treating $\mathbf{V}$ simply as a vector variable.

2. Show that div(curl $\mathbf{F}$) = 0, and curl grad ϕ = 0, identically.

3. When a viscous fluid flows steadily through a circular cylindrical pipe (Poiseuille flow), axis $z = 0$, its velocity field is given by
   $\mathbf{v} = (0, 0, v_z = (\Delta p/4\eta^2 l)(r^2 - R^2))$
   where Δp, η, R, l are independent of $r^2 = x^2 + y^2$.
   Calculate
   (*i*) curl $\mathbf{v}$;
   (*ii*) $\int_C \mathbf{v} \cdot d\mathbf{r}$ where C denotes the path $(x = 0, z = 0) \rightarrow (x = 0, z = 1) \rightarrow$
       $(x = R, z = 1) \rightarrow (x = R, z = 0) \rightarrow (x = 0, z = 0)$, all in the plane $y = 0$;
   (*iii*) use the results to verify Stokes' Theorem.

4. Consider a vector field which satisfies
   $$F_z = 0, \quad \frac{\partial F_y}{\partial x} = \frac{\partial F_x}{\partial y}, \quad \frac{\partial F_y}{\partial z} = \frac{\partial F_x}{\partial z} = 0.$$
   Show that the general theory of the last few chapters stipulates that $\mathbf{F}$ can be related to a potential function.
   Show that $\mathbf{F} = (-\partial\phi/\partial x, -\partial\phi/\partial y, 0)$ satisfies these equations and relate this to your previous discussion.

5. The internal magnetic field $\mathbf{h}(\mathbf{r})$ of a superconductor satisfies the London equation
   $\lambda^2 \text{curl}(\mu_0 \mathbf{J}) + \mathbf{h} = 0$
   where $\mathbf{J}(\mathbf{r})$ is the internal electric current density and λ is a constant. Combining this with the Maxwell equation (34.10) (with $\mathbf{B}$ replaced by $\mathbf{h}$) and assuming div $\mathbf{h} = 0$, use the identity of exercise 1 to show that $\mathbf{h}$ satisfies the equation
   $\mathbf{V}^2 \mathbf{h} = \lambda^{-2} \mathbf{h}$.
   Show also that $\mathbf{h} = (h(z), 0, 0)$ is a possible field and find a solution which satisfies the boundary condition, $h(0) = H$, $h(z \rightarrow \infty) = 0$.

6. A dry cell consists of a long cylindrical shell anode of radius R enclosing a thin rod cathode at its (z) axis. In the steady state a radially symmetric current density $\mathbf{J}(\mathbf{r})$ flows from cathode to anode through the interior

electrolyte. Show that $\mathbf{J} = \alpha\mathbf{r}/r^2$ satisfies the divergence equation, $\nabla\cdot\mathbf{J} = 0$, (the rod is assumed to exclude $\mathbf{r} = 0$) where $\mathbf{r}$ is the radius vector $(x, y, 0)$, and that the total cathode/anode current/unit length is $2\pi\alpha$.

Check that the magnetic field distribution $\mathbf{B}(\mathbf{r}) = \mu_0\alpha(yz/r^2, -xz/r^2, 0)$ satisfies the basic equations (i) curl $\mathbf{B} = \mu_0\mathbf{J}$; (ii) div $\mathbf{B} = 0$. By direct evaluation, verify Stokes' theorem, $\int_C \mathbf{B}\cdot\mathbf{dr} = \int_S$ curl $\mathbf{B}\cdot\mathbf{dS}$, where the loop C is a quadrant of the anode surface of height h (take it to extend from $x = R$, $y = 0$, to $x = 0$, $y = R$ in the plane $z = 0$, and completed in the reversed direction in the plane $z = h$). If $z = 0$ is chosen at the mid-plane of the cell, sketch the $\mathbf{B}$ field for a sequence of z planes and discuss its behaviour if the cell is viewed along positive and negative z-directions.

35
Conclusion

For the last hundred years, much of the theory of physical science has been concerned with the analysis of solutions of partial differential equations – the Laplace Equation, the Poisson Equation, Maxwell's Equations, the Wave Equation, the Diffusion Equation, Schrödinger's Equation....

We have touched upon some of these but have stopped at the very brink of a wider discussion of them, in their three-dimensional forms. We can already see or anticipate some of the lines of attack which might be used when the solution of a partial differential equation is not obvious.

First, the solution may be represented as a sum over some convenient set of functions and an attempt made to find the coefficients. This reduces the problem to one of matrix algebra, which is often a great relief. A further simplification of the problem is achieved if an awkward term can be discarded from the equation and the solution obtained without it. The missing term can then be put back in an approximate way by use of perturbation theory, a technique which is used everywhere in quantum mechanics but has its antecedents in early work on the dynamics of planetary orbits. Such methods have been given an elegant formal basis in the modern theory of linear operators.

Finally, the solution function may be represented by values on a discrete grid of points, whereupon the differential equation becomes a difference equation and we are once more in the realm of matrix algebra. In this way, the applied scientist can produce computer solutions to problems much too awkward for the more elegant methods.

35.1. Molecular dynamics simulation of the motion of atoms at the surface of a melting crystal. Trajectories of individual atoms are shown.

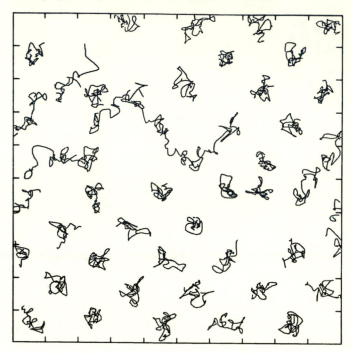

35.2. Equilibrium structure of a soap froth between two plates, simulated by a computer calculation. Note the periodic boundary conditions – the pattern matches on opposite sides. This is done to reduce the effects of finite size.

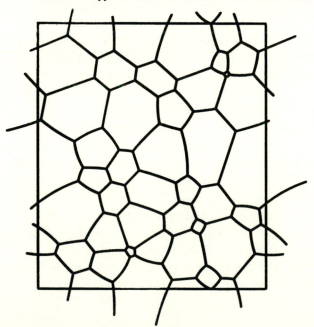

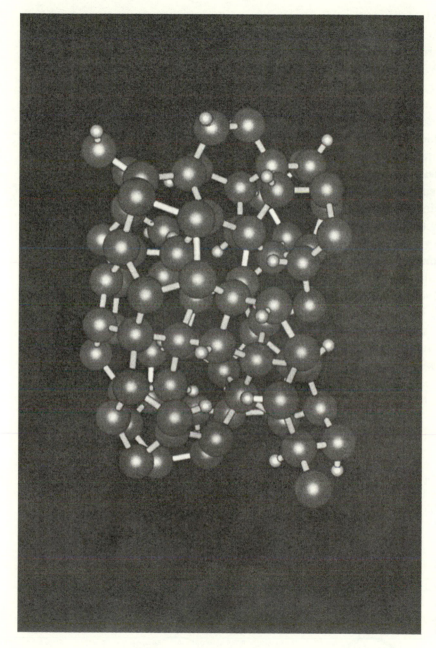

35.3. A splendid example of the use of modern computer graphical techniques to represent a three-dimensional structure which is the output of a calculation (Lawrence Livermore Laboratory).

All of this is the stuff of several years of study, but the underlying spirit is to be found in the more elementary topics covered in this book – *linear approximations, Fourier series, convergence of series* ....

At the same time, some areas of physical theory have been moving away from the traditional pre-occupation with partial differential equations. An exciting example is molecular dynamics, which seeks to model directly the motion of individual atoms or molecules under the influence of intermolecular forces. For example, fig. 35.1 shows the computed motion of atoms near the surface of a melting crystal. This must be a very old idea but its successful implementation is relatively recent, since it requires the largest of today's computers to make really useful calculations. This kind of theory is what engineers call 'simulation' and has been regarded by many physical scientists with a certain disdain, the suggestion being that no real understanding can come from its merely numerical output. There is some truth in this, but interesting numerical results usually find a neat expression in analytical form after due consideration. Also, the rapid development of computer graphics will make this kind of calculation more attractive and digestible. Further examples are shown in figs. 35.2, 35.3. At the time of writing, the latest graphical techniques are still having more impact on science fiction movies than science courses, but this must surely change soon.

35.4. The progress of physical science involves the interaction of each of the activities shown here. The emphasis of the present book is indicated by the weight of the lines.

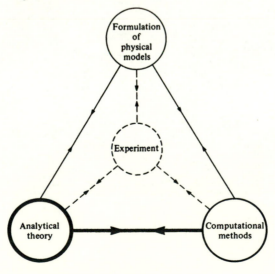

We leave the reader with a diagram which will remind him that all of this is part of a wider subject (fig. 35.4), no part of which should be studied in isolation. Sommerfeld once said 'If you want to be a physicist, you must do three things – first, study mathematics, second, study mathematics and third, do the same'. His achievements must lend credence to his advice, but most of us would do well to take a wider view.

EXERCISES

1. Write an essay based on fig. 35.3, discussing the relationships depicted.
2. Describe the various uses of the word 'vector' in this book.
3. Make a detailed study of the dynamics of the simple pendulum, for large as well as small amplitudes, in the spirit of fig. 1.1

36

Miscellaneous exercises

1. The Airy equation

$$y'' = xy$$

has solutions which are known as Airy functions. They describe, among other things, diffraction effects in optics and the quantum states of an electron in a metal-oxide-semiconductor transistor. Figure 36.1 shows

36.1. Numerical integration of the Airy Equation, using the Euler Method with $\Delta x = 0.01$

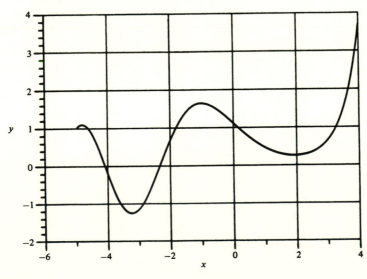

a numerical solution using Euler's method, starting from $y = y' = 1$ at $x = -5$ and using the step-length 0.01.

(a) Based on your experience with other calculations of this kind, roughly how accurate would you expect the calculation to be? (If you consult tables of Airy functions you may be able to check this.)

(b) *Sketch* the expected form of the solution for $x < -5$.

(c) In the limit $x \to \infty$, there are three possibilities: $y \to \infty$; $y \to -\infty$; $y \to 0$; depending on the initial conditions. If we try to reproduce the second type of behaviour by a numerical calculation of this sort, by choosing the correct initial conditions (which are quite close to the above ones), the solution will still eventually 'blow up', in practice. Why?

2. Find the terms in the series solution of Airy's equation about $x = 0$, up to x^5, starting from $y(1) = y'(1) = 1$. *Sketch* the form of the approximate solution given by these terms and indicate *roughly* the range in which it may be expected to be accurate to within 10%.

3. A ball of mass 1 kg is projected upwards at an angle of 45° with speed 1 ms^{-1} from a horizontal plane along which it bounces. At each bounce, it rebounds with a vertical component of velocity of magnitude half of that before impact. Sketch the resulting motion. By applying Newton's second law to the horizontal and vertical components of the motion, describe it quantitatively. In particular, what functions give the dependence of the maximum height between bounces on (a) number of bounces, (b) horizontal distance travelled at the point of maximum height?

4. In 1981, a school girl wrote to the Irish Times complaining that all the books and atlases that she consulted gave different lengths for the Shannon River. Presumably unknown to her, a book by B. Mandelbrot, already a classic, was becoming popular at the same time and explained, among other things, that a river does *not* have a unique length. It depends on the resolution with which it is measured (the size of the smallest wiggle which is included) and does *not* tend to a finite limit as the measurement is made more and more painstakingly. Such a strange curve is called a *fractal*; once it was no more than a mathematician's dream, soon it may be commonplace. Can you invent any simple geometrical construction which creates an anomalous curve of this sort?

5. In the last year or two, fast electronics and lately, opto-electronics, have revealed behaviour undreamt of by the early founders of function theory. It can arise in feedback loops when an optical beam is split, one point being delayed in phase and later recombined, allowing for optical interference. The resultant output signal, though in principle predictable by equations which contain no random inputs, shows a random functional behaviour, now known as *chaos*.

It is amusing (and instructive) for the student with a micro-computer to investigate a simpler but genuine instance of chaos. Consider the sequence

of quantities $x_1, x_2, x_3, \ldots x_n$ satisfying the feedback (difference) equation,
$x_{n+1} = 4\lambda x_n(1 - x_n)$.
Here λ is a 'control' parameter which can be assigned any value. A
discussion of the bizarre properties of the sequence of x_n is to be found in
an article by D.R. Hofstadter in *Scientific American* (1982). The author
notes that there is a critical value for λ above which the sequence is chaotic.
This is $\lambda = 0.892486417967\ldots$. With a starting value for x_1, say $x = 0.04$,
and with λ greater than the quoted value, his advice is to let the computer
run and 'hold on to your hat'.

    Although this is in some ways a very new subject, it is related to the
classic problem of *turbulence*.

6. Examine the magnitude of the error associated with the approximation
which leads to simple harmonic motion for the pendulum (chapter 20), as
follows.

    Find the form (to lowest order) of the terms which were neglected in the
equation (20.2). Use the approximate simple harmonic solution to
compare the magnitudes of these terms with those which are retained.

7. The error function $\text{erf}(x) = 2/\pi^{\frac{1}{2}}\int_0^x \exp(-t^2)\,dt$ is useful, since it is the
integral of the normal distribution (chapter 2). The integral cannot be
performed in terms of the usual functions and rules of integration, hence
the definition of this new function.
Draw up a table of values of $\text{erf}(x)$ for $x = 0.5, 1, 1.5, 2.0$. Make an estimate
of $\text{erf}(\infty)$ from this table and consult any large set of mathematical tables
to confirm this value.

8. The Fermi-Dirac function
$f(E) = [1 + \exp\beta(E - E_0)]^{-1}$
is used to describe the probability of occupation of (quantum-mechanical)
electron states of energy E in metals and semiconductors. *Sketch* this
function, showing how it varies as $\beta \to \infty$. (Physically, this means the
low temperature limit.) For what range of E can f be approximated by
$\exp[-\beta(E - E_0)]$ to within 1%? This is a standard approximation in
semiconductor physics.

9. Write a short essay on the various related meanings of *linearity* and its
significance in different fields of mathematics.

10. Write a short essay, with suitable examples, based on the diagram shown
below (Fig 36.2).
(From *The Mathematical Experience* by P.J. Davis and Heirsh, Houghton
Mifflin Co., Boston, 1981, p. 129.)

12. (*a*) The three-dimensional form of the wave equation is
$$\nabla^2\psi \equiv \frac{\partial^2\psi}{\partial x^2} + \frac{\partial^2\psi}{\partial y^2} + \frac{\partial^2\psi}{\partial z^2} = v^{-2}\frac{\partial^2\psi}{\partial t^2}.$$

Show that a solution, analogous to (26.3), is

$$\psi = A \sin(\mathbf{k} \cdot \mathbf{r} - \omega t)$$

where $\mathbf{r} = (x, y, z)$, and the vector $\mathbf{k}$ and scalar ω are constants. Describe its qualitative form. The vector $\mathbf{k}$ is called the *wave vector*.

(b) Whenever a wave disturbance emanates from a point source, it is useful to re-express the wave equation in terms of spherical polar coordinates (radius r, angles θ, ϕ see exercise 5, chapter 11). A spherically symmetric solution $\psi(r, t)$ must satisfy

$$r^{-2} \frac{\partial}{\partial r} \left(r^2 \frac{\partial \psi}{\partial r} \right) = v^{-2} \frac{\partial^2 \psi}{\partial t^2}$$

Verify that a solution is

$$\psi = A r^{-1} \cos[k(r + vt) + \phi].$$

Describe its qualitative form, and discuss how this is related *locally* to the solution in part (a).

13. The method of *separation of variables* provides a derivation of the solutions (26.6) of the wave equation. It proceeds by assuming the form $\psi(x, t) = F(x)G(t)$ and substituting this in the wave equation given by (26.6), (26.7). The resulting equation may be written

$$\frac{1}{F} \frac{d^2 F}{dx^2} = v^{-2} \frac{1}{G} \frac{d^2 G}{dt^2}.$$

The left-hand side does not depend on t, while the right-hand side does not depend on x – therefore each can be set equal to the same *constant*. This gives two *ordinary* differential equations to be solved for F and G. Complete the derivation of (26.6) in this way.

14. In some crystal structures, such as those of most solid elements, the

36.2. The mysterious relation between mathematics and the physical world.

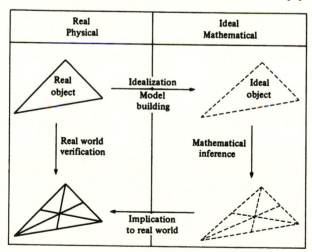

position of every atom may be specified by a vector

$\mathbf{r} = l\mathbf{a} + m\mathbf{b} + n\mathbf{c}$

where l, m, n are integers and $\mathbf{a}, \mathbf{b}, \mathbf{c}$ are 'primitive vectors'.

Draw a sketch to illustrate this, including the parallelepiped of which $\mathbf{a}, \mathbf{b}$ and $\mathbf{c}$ are three edges. Show that the atomic volume (volume per atom) is given by the volume Ω of this parallelepiped. A useful formula for this is $\Omega = |\mathbf{a} \cdot \mathbf{b} \times \mathbf{c}|$. Give a derivation of this formula by relating it to the geometrical formula *Volume = Base × Height*.

15. Sketch the functions of which the delta function $\delta(x)$ may be regarded as the first and second derivative, respectively.

16. The integral

$$I = \int_0^\infty \frac{x^3 \, dx}{e^x - 1}$$

crops up in the theory of radiation, in deriving Stefan's Law from Planck's Law.

   (*a*) Make a rough ('back-of-envelope') estimate of I.

   (*b*) Make a more accurate numerical estimate by any method.

   (*c*) By expanding the factor $(1 - e^{-x})^{-1}$ as a power series, relate I to the following series

$$s = 1^{-4} + 2^{-4} + 3^{-4} \ldots,$$

and use this to make a still more accurate estimate of I.

   (*d*) There is in fact an exact formula for I. Consult a physics textbook to find this, and compare it with your own results.

17. Give a proof of the following relation between averages taken with respect to a distribution $f(x)$

$$\langle (x - \langle x \rangle)^2 \rangle = \langle x^2 \rangle - \langle x \rangle^2.$$

Students who have studied moments of inertia may notice a resemblance to the Parallel Axis Theorem. If so, comment.

18. The power output of a solar cell under constant illumination is given by the product of current I and voltage V, where these are related by

$I = I_s - A(\exp(BV) - 1)$

(cf. exercise 10, chapter 10).

In practice the particular I, V values depend on the total resistance in the circuit to which the cell is connected. What is the *maximum* possible output?

19. If we suppose that the period T of a pendulum depends only on its length l and the gravitational constant g, it follows from 'dimensional arguments' that

$$T \propto \sqrt{\frac{l}{g}}$$

since both sides have the dimension of *time*. In chapter 20 this is derived more directly and the constant of proportionality is found to be 2π. This enables the pendulum to be used to measure g. However the derivation (and the above assumption) is only valid *in the limit of small amplitude oscillations*. What can be said, on dimensional grounds alone, about the case of finite amplitude? Would you expect finite amplitude effects to contribute to the systematic error

(a) when the pendulum is used to measure g,

(b) when the pendulum is used to compare g at two different points, using the above formula?

20. A sphere of density ρ and radius a falling in liquid of density ρ_0 with velocity v is subject to the following forces: Gravity (corrected for buoyancy): $\frac{4}{3}\pi a^3(\rho - \rho_0)g$, Viscous drag: $6\pi\eta va$ (η = coefficient of viscosity) Set up the equation which governs the motion of the falling sphere. Solve it and find the time necessary for the sphere to reach 90% of its terminal (limiting) velocity, starting from rest, as a formula involving ρ, a, η, etc.

21. Write a simple computer program to find the minimum of a function of three variables, given that it lies within a prescribed 'box'. Use the following strategy:

(a) *Random search*. This alone will suffice but is clearly inefficient.

(b) *Refinement* of the best estimate of (a), using the gradient operator in an iterative manner.

Why is (a) to be recommended as a preamble to (b), rather than starting from a single arbitrary point?

Note: *Variational* methods formulate physical problems in such a way that the solution can be found by such a process. However, when the function happens to be (or is restricted to be) *quadratic*, matrix methods provide direct, efficient solutions. More general minimisation or 'optimisation' methods are of great interest in practical engineering, for obvious reasons.

22. In quantum mechanics, Schrödinger's equation for the wave function of an electron with energy E in a potential well $U = \frac{1}{2}qx^2$ is the eigenvalue equation

$$\frac{d^2\psi}{dx^2} - \left(\frac{4\pi^2 mq}{h^2}\right)x^2\psi = \frac{-2mE}{h^2}\psi$$

where m is the mass of the electron and h is Planck's constant. In what ranges should the solution have an oscillatory/exponential form? Show that a Gaussian function of x gives an eigenfunction. Physically acceptable eigenfunctions must go to zero at $\pm\infty$. Can you *sketch* other ones, corresponding to higher eigenvalues E?

23. The velocity of water waves in deep water is given as a function of wavelength λ by

$$v = (g\lambda/2\pi + 2\pi T/\rho\lambda)^{\frac{1}{2}}$$

The constants g, T and ρ are the acceleration due to gravity, the surface tension and the density of water.

Sketch this function and derive formulae for the minimum velocity and the corresponding wavelength. If a combination of two sinusoidal water waves propagates with a profile which does not change, what must be the relation between the two wavelengths?

24. A standing wave is set up in a stretched string. How would you expect the total *energy* associated with this wave to depend on the physical parameters characterising the string and the standing wave? (Use dimensional and physical arguments to develop the formula.)

25. Consider what happens locally in a solid under stress. Relative to a given point (fixed in the solid itself), points close to it are displaced from their original positions. To a first approximation we can write a linear relation
$$\Delta \mathbf{r} = \mathbf{M} \mathbf{r}$$
for this displacement, where $\mathbf{M}$ is a matrix.

    Suppose that at one point the matrix is
$$\begin{bmatrix} 1 & 0 & 1 \\ 0 & 1 & 1 \\ 2 & 2 & 2 \end{bmatrix}.$$

    Express this as the sum of a symmetric and an antisymmetric matrix. Give a detailed interpretation of the antisymmetric matrix, considered on its own (see chapters 8, 34). What then is the significance of the symmetric part?

26. Consider a long (effectively infinite) chain of atoms making up a molecule. The atoms are equally spaced by a distance d. They interact according to the Morse potential (chapter 12, exercise 10). Units are chosen such that the parameters V_0 and r_e in the potential $V(r)$ are both unity.

    The energy per atom is written as
$$U = 2 \sum_{n=1}^{\infty} V(nd).$$

    Justify this formula. Sum the series to obtain an explicit expression for U as a function of the interatomic separation d. Hence derive an equation for the value of d which minimises U. Solve this and compare the result with that for a diatomic molecule. Explain the difference between these results. (Hint: it is helpful to replace d by x at an early stage, where $x = \exp(-d/a)$.)

27. As an alternative to the axes used in exercise 10, of chapter 8, we can choose axes such that the positions of the four Cl atoms in CCl_4 are given by
$$\mathbf{r}_1 = (1, 1, 1), \quad \mathbf{r}_2 = (-1, -1, 1),$$
$$\mathbf{r}_3 = (1, -1, -1), \quad \mathbf{r}_4 = (-1, 1, -1),$$
in suitable units. Show that these vectors are indeed oriented at the 'tetrahedral angle' $\cos^{-1}(-\tfrac{1}{3})$, with respect to each other. Apply the

matrix

$$\mathbf{R} = \begin{pmatrix} 0 & 1 & 0 \\ 0 & 0 & 1 \\ 1 & 0 & 0 \end{pmatrix}$$

to each of these vectors and interpret your results, with the help of a careful *sketch*.

28. Give an argument, based on an analogy with vectors, justifying the inequality

$$\left(\int_a^b fg \, dx \right)^2 \leqslant \left(\int_a^b f^2 \, dx \right) \left(\int_a^b g^2 \, dx \right).$$

Use this to derive an *upper bound* for the Fourier coefficients of a function $f(t)$.

29. A theoretical formula for the total energy of a KCl crystal, relative to free K^+ and Cl^- ions, consists of two terms (R is the nearest neighbour distance in nm):

Short range repulsion :	$0.6 \times 10^4 \exp(-R/0.030)$ eV/atom
Coulomb interaction : of ions	$0.63 \, R^{-1}$ eV/atom.

Sketch the total energy as a function of R. Find the position of the minimum R_0 to within 0.1% accuracy by any method.

30. (*The Ehrenfest game*, useful in thinking about the relation between kinetic theory and thermodynamics.) One hundred balls, numbered individually 1–100, are placed in container A. Container B is initially empty. Random numbers (1–100) are selected and the corresponding ball is transferred to the *other* container (A to B or B to A, as the case may be) each time.

   (a) Make a computer experiment to study the variation of the number of balls N in container A as a function of the total number of such transfers, n, for a few hundred transfers. (This might also be done by recourse to a table of random numbers, without a computer, perhaps with somewhat fewer balls.) Repeat the experiment a few times and take the mean of your results for each value of n. Comment on the effect of this.

   (b) Can you think of an equation for the mean $\bar{N}(n)$, calculated from a large number of such trials, of the form:

   $\bar{N}(n+1) = \bar{N}(n) + \cdots ?$

   This is a *difference equation*. In this case it can be well approximated by a *differential* equation. The latter has a familiar solution, but the difference equation itself is just as easily solved. (Put $\bar{N} = x^n$, solve for x.) Find either solution and compare with your computed results.

31. Plot as functions of ω the real and imaginary parts of the complex dielectic constant ε defined by $\varepsilon(\omega) = \varepsilon_1(\omega) + i\varepsilon_2(\omega) = (\omega - \omega_0 + i\gamma)^{-1}$.
    Show that $\int_{-\infty}^{\infty} \varepsilon_2(\omega)d\omega$ is independent of γ.
    Here ω is the (circular) frequency of light and the complex refractive index $n = \varepsilon^2$ can be used to describe its propagation through a particular material. The above formula applies, for example, in the vicinity of an absorption line in a gas.

32. Rewrite the basic formula for a sinusoidal wave
    $$\psi = \cos(kx - \omega t)$$
    as the real part of a complex expression for ψ. What would be the effect of making k and/or ω *complex* in such an expression?

33. For light, complex ε and complex k in exercise 32 can be related by the following equation
    $$\omega^2 \varepsilon(\omega) = c^2 k^2$$
    where ω is (real) circular frequency and c is the velocity of light.
        Suppose ε_2 is small and ε_1 is *negative* (as in the case of a metal at low frequencies): what is the implication for k?

34. The two-dimensional version of the diffusion equation is
    $$\nabla^2 \phi = D\partial\phi/\partial t$$
    where $\nabla^2 = \partial^2/\partial x^2 + \partial^2/\partial y^2$.
    In part of a nuclear reactor, the density $n(r)$ of neutrons is independent of the vertical coordinate z, so that it obeys this equation. A numerical simulation of the reactor is set up, using values of n on a *triangular* grid of points in the (x, y) plane. What expression should be used for ∇^2, in terms of these values? (Triangular grids are often used in two-dimensional problems, in preference to the more obvious square grid.)

35. How may the formula derived in exercise 9 of chapter 27 be
    (*i*) simplified for evaluation of components,
    (*ii*) expressed as a matrix relation between **a** and **r**?

36. According to Kleiber's law, the height h of a tree is proportional to the diameter of its trunk *to the $\frac{2}{3}$ power*. Like many other bits of biological folklore, this is derivable from physical principles! If the trees of a plantation have a diameter of 0.5 m and a height of 20 m, and are observed to be increasing in diameter at a rate of 0.06 m per year, what is the corresponding gain in height, as predicted by Kleiber's law?

37. A simple pendulum consists of a mass attached to the lower end of a light rod of length l and freely hinged at its upper end. It can swing freely about the vertical direction with a frequency $(l/g)^{\frac{1}{2}}$. In the so-called 'vertical pendulum' the hinge end is forced to vibrate rapidly in a vertical direction with an amplitude $a \ll l$, and at a frequency $\omega \gg (l/g)^{\frac{1}{2}}$. As a result the pendulum is now able to swing about the vertical direction but in an *inverted* position, i.e. with the mass *above* the hinge. Given that the

appropriate differential equation describing small angular oscillations $\theta(t)$ is

$$\frac{d^2\theta}{dt^2} + \left(\frac{a\omega^2}{l}\cos\omega t - \frac{l}{g}\right)\theta = 0$$

check that an approximate solution (which neglects terms of the order of $l/g\omega^2$) is $\theta = A\cos\Omega t(1 + a/l\cos\omega t)$,

where $\Omega^2 = \frac{1}{2}(a\omega/l)^2 - l/g$.

(During one oscillation, there are many hinge vibrations so that $\cos^2\omega t$ may be replaced by its average, namely $\frac{1}{2}$.)

38. Use a random number generator or table (or a coin!) to make the following experiment.

   A coin is tossed repeatedly until the sequence 'heads' – 'heads' occurs. The number of tosses is noted and this process is repeated many times, leading to an estimate of the *average* number of tosses needed to generate this sequence. The same is done for 'heads' – 'tails'. Most people are startled to learn that the two averages are *different*. This shows that common sense must not be pushed too far in dealing with probabilities. The mathematical theory of probability (based on the same rules which we have used) must not be postponed too long!

39. When visualised in terms of fluid flow rather than a magnetic field, the field (34.11) seems paradoxical. Away from $\mathbf{r} = 0$, it is *irrotational* yet is seems obvious that a small element of fluid circling the origin *must* rotate through 2π! Discuss....

40. Use a random number generator or table to specify random numbers within the square whose corners are $(0,0)$, $(0,1)$, $(1,0)$, $(1,1)$. By working out the fraction of these numbers that lie within a circle of radius 0.5 centred at $(0.5, 0.5)$, make an estimate of the number π.

41. In chapter 6, a vector quantity $\Delta\phi$ was introduced to represent a *small* rotation, with the warning that it was not to be considered as an increment of some function $\phi(t)$. Why not? After all, if a body is moving with one point fixed, we may certainly relate its position at time t to that at time 0 by a finite rotation and this may be represented by ϕ. Why is this not a good idea? Where else in this book does a similar difficulty occur?

Index

Lightning Source UK Ltd.
Milton Keynes UK
UKOW051914011211

183045UK00001B/153/P